Edexcel
advancing learning, changing lives

D1321624

Series Editors
Trevor Johnson & Tony Clough

Homework Book

Edexcel
GCSE Mathematics

Higher Tier
Linear Course

Published by: Edexcel Limited, One90, High Holborn, London WC1V 7BH

Distributed by: Pearson Education Limited, Edinburgh Gate, Harlow, Essex, CM20 2JE, England
www.longman.co.uk

First published 2006

Fourth impression 2008

ISBN 978-1-903-13391-0 ✓ 𝒥𝒢
Cover design by Juice Creative Ltd.

Typeset by Tech-Set, Gateshead

Printed in Malaysia (CTP-VVP)

The publisher's policy is to use paper manufactured from sustainable forests.

Every effort has been made to trace the copyright holders and we apologise in advance for any unintentional omissions. We would be pleased to insert the appropriate acknowledgement in any subsequent edition of this publication.

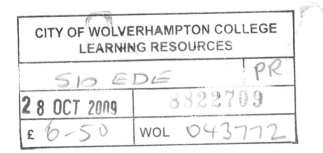

Contents

Homework

Chapter 1 Number

Exercise 1A

1 Which of the following numbers are factors of 24?

   a 12   **b** 6   **c** 9

   d 4   **e** 3   **f** 24

2 Find all the factors of 40. Write them in factor pairs.

3 List all the factors of

   a 18   **b** 20   **c** 42

   d 54   **e** 88   **f** 96

4 List all the common factors of

   a 8 and 10   **b** 9 and 12

   c 24 and 36   **d** 36 and 48

5 a Write down the first 3 multiples of 7

   b Write down the first multiple of 23 that is greater than 100

   c How many multiples of 9 are less than 100?

   d How many multiples of 8 are between 100 and 200 inclusive?

6 $y = x^2 + x + 11$

   a Write down the value of y **i** when $x = 0$, **ii** when $x = 1$

   b Find a whole number value of x for which y is **not** prime.

7 p and q are different prime numbers.

   a Write down all the factors of pq.

   p, q and r are different prime numbers

   b Write down all the factors of pqr.

   A number is the product of n different prime numbers.

   c How many factors will it have?

8 Find all the factors of 1419

9 a Show that 17 is a factor of 2431.

   b Find all the factors of 2431 that are less than 20

10 Find 2 prime numbers between 120 and 140.

Exercise 1B

1 Work out

   a $(+2) \times (-5)$ **b** $(-3) \times (-5)$

   c $(+4) \times (+3)$ **d** $(-6) \times (-3)$

2 Work out

   a $(+8) \div (-4)$ **b** $(+18) \div (-3)$

   c $(+28) \div (+7)$ **d** $(-20) \div (-5)$

3 Find the missing directed number

   a $(+6) \times (\ ) = (-18)$

   b $(+16) \div (\ ) = (-2)$

   c $(\ ) \times (-5) = (+30)$

   d $(\ ) \times (+6) = (-36)$

   e $(+28) \div (\ ) = (+4)$

4 Work out the product of

   a (-4) and $(+4)$ **b** (-5) and $(+6)$

   c $(+4)$ and $(+7)$ **d** (-6) and (-7)

5 a Multiply $(+8)$ by (-2). Divide your answer by (-4)

   b Multiply (-4) by (-6). Divide your answer by $(+3)$

   c When the product of (-8) and $(+4)$ is divided by a number the answer is (-4). What is the number?

Exercise 1C

1 Work out

   a 7^2 **b** 11^3 **c** $(-6)^2$

2 Work out

   a 100^2 **b** 100^3 **c** $\sqrt{4^2}$

   d $\sqrt{4 \times 16}$ **e** $\sqrt[3]{8^2}$ **f** $\sqrt[3]{7^3}$

3 Work out

a $\sqrt{9} \times 4^2$ **b** $2^4 + 3^3$

c $\sqrt[3]{-27} + 3^2$ **d** $(-2)^3 + 3^3 - (-4)^3$

e $\sqrt[3]{125} \times \sqrt{100}$ **f** $\dfrac{12^2}{3^2}$

Exercise 1D

1 Write as a power of 3

a $3^3 \times 3^2$ **b** $3^8 \div 3^2$ **c** $3^7 \div 3^4$

d $3^9 \div 3^5$ **e** $3^6 \times 3^2$ **f** 3×3^4

2 Write as a power of a single number.

a $4^5 \div 4^2$ **b** $3^5 \times 3^4$

c $6^4 \times 6^6$ **d** $7^4 \times 7^8 \div 7^5$

3 Write as a power of a single number.

a $\dfrac{4^8 \times 4^2}{4^7}$ **b** $\dfrac{5^3 \times 5^5}{5^6}$ **c** $\dfrac{3^{10}}{3^4 \times 3^2}$

d $(4^2)^3$ **e** $\dfrac{7^4 \times 7^6}{7^2 \times 7^5}$ **f** $\dfrac{8^3 \times 8^5}{8}$

4 Find the value of each of the following.

a $3^6 \div 3^4$ **b** $\dfrac{5^4 \times 5}{5^6}$ **c** $\dfrac{(3^2)^3}{3^5}$

5 Find the value of n in each of the following.

a $24 = 3 \times 2^n$ **b** $32 = 2^n$

c $56 = 7 \times 2^n$ **d** $72 \div 2^n = 3^2$

6 Find the value of k in each of the following

a $\dfrac{2^k \times 2^4}{2^6} = 2^3$ **b** $\dfrac{5^3 \times 5^k}{5^5} = 5^4$

c $\dfrac{3 \times 3^4}{3^k} = 3^2$ **d** $\dfrac{7^3 \times 7^4}{7^5} = 7 \times 7^k$

Exercise 1E

1 Work out

a $12 \div 2^2$ **b** $36 \div 3^2$

c $8 - 2^2$ **d** 12×2^2

e $2^3 \times 3$ **f** 9×1^2

g $8^2 - 2^2$ **h** $(-2)^3 - (-3)^3$

2 Work out

a $5 - 4^2$ **b** $7 - 3^2$

c $(14 \div 2)^2$ **d** $2^3 - 3^2$

e $(2 \times 4)^2$ **f** $(-2 - 4)^3$

3 Work out

a $6^2 - 3^2$ **b** $7^2 - 3^2$

c $3^3 + 3^2 + 3$ **d** $(2 \times 5)^3$

e $64 \div 4^2$ **f** $6^2 - (3 + 1)^2$

4 Work out

a $\sqrt{49} \times 2$ **b** $\sqrt{16 \times 4}$

c $2^3 \div \sqrt{16}$ **d** $\sqrt{36} \times 2^3$

e $\sqrt{25} \times 5$ **f** $3^3 \div \sqrt{9}$

5 Work out

a $\sqrt{36} \times \sqrt{4}$ **b** $\sqrt{36} + \sqrt{4}$

c $\sqrt{100} \div \sqrt{4}$ **d** $\sqrt{25} \times \sqrt[3]{1000}$

Exercise 1F

1 Work out

a 7.2^2 **b** 16^2 **c** 0.52^2

2 Work out

a $23 + 13^2$ **b** $640 - 24^2$

c $49^2 + 32^2$ **d** $3.6^2 - 3.5^2$

3 Work out

a 4.1^3 **b** 11^3

c $21^3 - 20^3$ **d** $1000 - 8^3$

4 Find the square root of each of these numbers

a 676 **b** 4096 **c** 5.29

5 Work out each of the following. Give your answers correct to 1 decimal place.

a 6.3^2 **b** $\sqrt{20}$

c $2.5 \times \sqrt{10}$ **d** $3.1^2 + \sqrt{20.5}$

e $\dfrac{1.2^2 + 5.6}{2.7}$ **f** $\dfrac{6.4}{2.8 \times 1.7^2}$

6 Work out the following. Give your answers correct to 1 decimal place.

a $\dfrac{5.8 \times 4.7^2}{2.3 \times 4.5}$ b $\dfrac{3.8^3 + 14.7}{6.4}$

c $\dfrac{12.8 + 3.6 \times \sqrt{24}}{\sqrt{20} + \sqrt{30}}$ d $16.4 + \dfrac{52.7}{2.5^2}$

7 Find the reciprocal of each of the following numbers

a 5   b 20   c 1.25

Exercise 1G

1 Find the three prime numbers between 40 and 50.

2 Find two prime numbers which are factors of 50

3 Find two prime numbers which have a difference of 2.

4 Write the following numbers as a product of two prime factors

a 14   b 26   c 51

5 Find the Highest Common Factor (HCF) of the following pairs of numbers.

a 16, 20   b 18, 24   c 20, 24
d 20, 30   e 24, 36   f 36, 42

6 Find the Lowest Common Multiple (LCM) of the following pairs of numbers.

a 10, 15   b 12, 15   c 12, 16
d 9, 12   e 12, 18   f 10, 16

7 Find the Lowest Common Multiple (LCM) of the following triples of numbers

a 10, 15, 20 b 16, 24, 32
c 18, 24, 42 d 24, 32, 48

8 p is a prime number greater than 10.

a Find the Highest Common Factor (HCF) of $7p$ and $5p$.

b Find the Lowest Common Multiple (LCM) of $7p$ and $5p$.

9 Express each of the following as a product of powers of its prime factors.

a 28   b 40   c 60
d 84   e 96   f 108

10 Written as products of their prime factors $x = 2^3 \times 3^2 \times 5$, $y = 2 \times 3^3 \times 7$

a Find the Highest common factor (HCF) of x and y.

b Find the Lowest Common Multiple (LCM) of x and y.

11 Find all positive integer values of n less than or equal to 10 for which $2^n + 1$ is a prime number.

12 Write each of these numbers as a product of powers of its prime factors.

a 286   b 1615   c 455
d 646   e 363   f 1617

13 Find the lowest common multiple of these pairs of numbers.

a 27 and 33   b 30 and 36
c 42 and 48   d 84 and 96

Chapter 2 Angles

Exercise 2A

In Questions **1–6**, find the size of each of the angles marked with letters and show your working. The diagrams are not accurately drawn.

1

2

3

4

5

6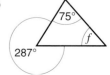

7

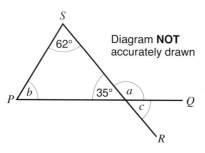

Diagram **NOT** accurately drawn

a Work out the value of x.

b Work out the value of y.

(1388 January 2004)

8

Diagram **NOT** accurately drawn

In the diagram, PQ and RS are straight lines.

a i Work out the value of a.
ii Give a reason for your answer.
b i Work out the value of b.
ii Give a reason for your answer.
c i Work out the value of c.
ii Give a reason for your answer.

(1385 June 1999)

Exercise 2B

In Questions **1–10**, find the size of each of the angles marked with letters and show your working. The diagrams are not accurately drawn.

1

2

3

4

5

6

7

8

9

10

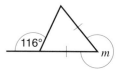

Exercise 2C

In Questions **1–6**, find the size of each of the angles marked with letters and show your working. The diagrams are not accurately drawn.

1

2

3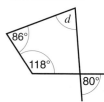

4

5

6

7 The diagram shows an isosceles trapezium.

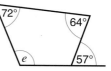

Diagram **NOT** accurately drawn

Work out the value of n.

8 The diagram shows a kite $ABCD$.

Diagram **NOT** accurately drawn

a i Write down the value of x.
ii Give a reason for your answer.
b i Work out the value of y.
ii Give a reason for your answer.
c Show that ABC is an equilateral triangle.

(4400 May 2005)

Exercise 2D

In this exercise, the diagrams are not accurately drawn.

1 Find the sum of the angles of a 14-sided polygon.

2 Find the sum of the angles of a 30-sided polygon.

In Questions **3** and **4**, find the size of each of the angles marked with letters and show your working.

3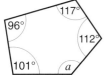

4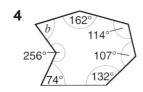

5 Work out the size of each interior angle of a regular 18-sided polygon.

6 Work out the size of the angle at the centre of a regular 30-sided polygon.

7 The angle at the centre of a regular polygon is 72°.
How many sides has the polygon?

8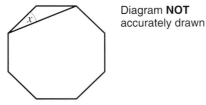

Diagram **NOT** accurately drawn

The diagram shows a regular octagon.
Work out the size of the angle marked x.

(1388 November 2005)

9 a Work out the size of the angle at the centre of a regular 15-sided polygon.

b Draw a circle with a radius of 5 cm and, using your answer to part **a**, draw a regular 15-sided polygon inside the circle.

10 The diagram shows a pentagon.

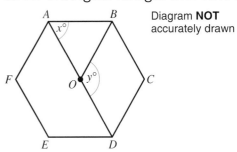

a Work out the size of each of the angles marked with letters.

b Work out $v + w + x + y + z$

11 $ABCDEF$ is a regular hexagon with centre O.

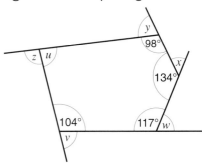

Diagram **NOT** accurately drawn

a What type of triangle is ABO?

b **i** Work out the size of the angle marked $x°$.

ii Work out the size of the angle marked $y°$.

c **i** What type of quadrilateral is $BCDO$?

ii Copy this diagram to show how three such quadrilaterals can tessellate to make a hexagon.

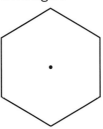

(1384 November 1996)

Exercise 2E

1 At a vertex of a polygon, the size of the interior angle is 107°.
Work out the size of the exterior angle.

2 The sizes of three of the exterior angles of a quadrilateral are 93°, 104° and 85°.
Work out the size of the other exterior angle.

3 The size of four of the exterior angles of a pentagon are 57°, 89°, 48° and 112°.
Work out the size of the other exterior angle.

4 Work out the size of each exterior angle of a regular 20-sided polygon.

5 For a regular 45-sided polygon, work out
  a the size of each exterior angle,
  b the size of each interior angle.

6 The size of each exterior angle of a regular polygon is 24°.
Work out the number of sides the polygon has.

7 The size of each interior angle of a regular polygon is 174°. Work out
  a the size of each exterior angle
  b the number of sides the polygon has.

8 Is it possible to have a regular polygon with 27° as the size of each of its exterior angles?
Explain your answer.

Exercise 2F

1
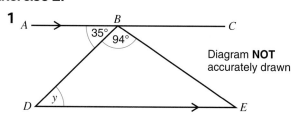

Find the size of the angle marked y.
Give a reason for your answer.

(1388 March 2002)

2
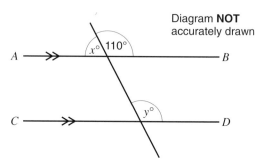

Diagram **NOT** accurately drawn

AB is parallel to CD.

  a Work out the size of angle $x°$.

  b **i** Work out the size of angle $y°$.
     ii Give a reason for your answer.

(1388 January 2002)

3

Diagram **NOT** accurately drawn

ABC is parallel to $DEFG$. $BE = EF$.
Angle $ABE = 38°$.

  a **i** Find the value of x.
     ii Give a reason for your answer.

  b Work out the value of y.

(1388 March 2005)

4
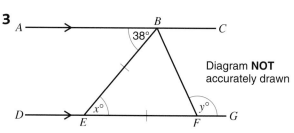

Diagram **NOT** accurately drawn

$AB = BC$. Angle $ACB = 63°$. ACE and BCD are straight lines.

  a **i** Find the size of the angle marked $p°$.
     ii Give a reason for your answer.

  b Work out the size of
     i the angle marked $m°$
     ii the angle marked $r°$.

AB is parallel to *DE*.

c **i** Find the size of the angle marked *t°*.

ii Explain how you worked out your answer.

(1385 November 1999)

5 In the diagram, *PQR* and *PST* are straight lines.
QS and *RT* are parallel lines.
Angle *QRT* = 70°. Angle *QST* = 120°.

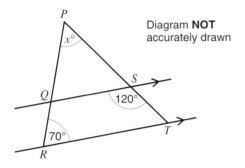

Diagram **NOT** accurately drawn

Work out the value of *x*.
Give a reason for each step in your working.

(4400 May 2005)

Exercise 2G

In this exercise, the diagrams are not accurately drawn.
Find the size of each of the angles marked with letters.

Give reasons for your answers.

1

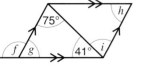

2

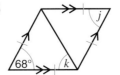

3

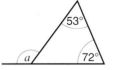

4

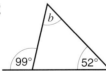

5

6

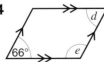

Exercise 2H

1 Measure the bearing of *B* from *A*.

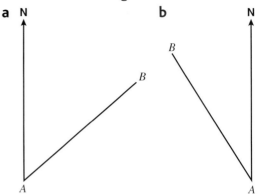

2 Draw diagrams similar to those in Question 1 to show the bearings

a 073° **b** 169° **c** 218°

d 270° **e** 329°

3 The diagram is part of a map showing the positions of several cities.

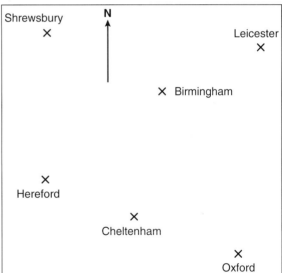

Copy the diagram. Measure and write down the bearing of

a Leicester from Cheltenham

b Oxford from Hereford

c Oxford from Birmingham

d Hereford from Shrewsbury

e Leicester from Oxford

f Cheltenham from Birmingham

g Shrewsbury from Leicester

h Birmingham from Oxford

i Hereford from Birmingham

j Shrewsbury from Cheltenham.

4 The diagram shows the position of two ships, A and B.

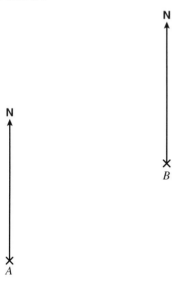

Copy the diagram.
A ship C is on a bearing of 036° from ship A.
Ship C is also on a bearing of 284° from ship B.
Draw an accurate diagram to find the position of ship C.
Mark the position of ship C with a cross **X**.
Label it C.

5 Work out the bearing of

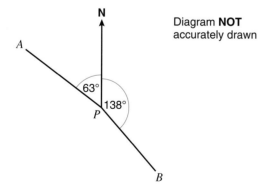

Diagram **NOT** accurately drawn

a A from P **b** P from B.

Chapter 3 Scatter graphs

Exercise 3A

1 The table shows the number of donkey rides at Blackpool beach and the number of hours of sunshine on each of 10 days.

Number of donkey rides	Number of hours of sunshine
130	3
170	3.5
185	4
210	5
220	5.5
250	6
260	7.5
310	8.5
325	9
345	9.5

a The first 6 points in the table have been plotted on the scatter graph.
Copy and complete the scatter graph.

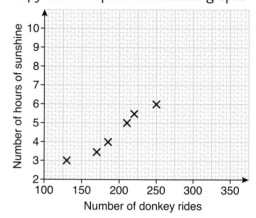

b Describe the relationship between the number of donkey rides and the number of hours of sunshine.

2 The table shows the hours of sunshine and the rainfall, in mm, in 10 towns during last summer.

Sunshine (hours)	Rainfall (mm)
650	10
455	20
560	15
430	29
620	24
400	28
640	14
375	30
520	25
620	20

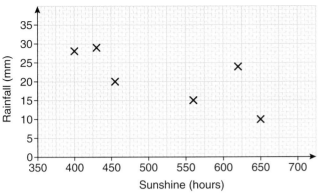

Sunshine (hours)

The points for the first six results in the table have been plotted on a scatter diagram.

a   Copy and complete the scatter diagram.
b   Describe the relationship between the hours of sunshine and the rainfall.

3   The table shows, for each month of last year, the average midday temperature, in °C, and the number of visitors, in thousands, to a country park.

Average midday temperature in °C	Number of visitors in thousands
3	6
6	6
9	8
11	9
12	12
18	16
21	20
23	20
19	18
15	13
8	7
4	4

a   Draw a scatter graph to show the information in the table.

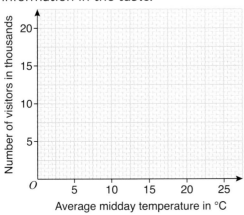

Average midday temperature in °C

b   Describe the relationship between the average midday temperature and the number of visitors.
c   Draw a line of best fit on the scatter graph.

Exercise 3B

1   10 students each took a French test and a German test. The table shows their marks.

French mark	German mark
44	48
30	35
40	45
50	54
14	18
20	22
32	36
34	38
20	25
45	50

The first 8 points in the table have been plotted on the scatter graph.

a   Copy and complete the scatter graph.

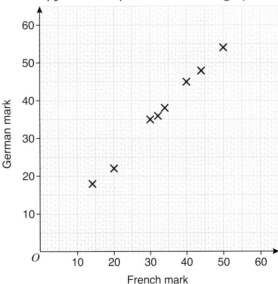

French mark

b   What type of correlation does this scatter graph show?
c   Draw a line of best fit on the scatter diagram.
d   Use your line of best fit to estimate
   i   the German mark for a student with a French mark of 26
   ii the French mark for a student with a German mark of 43

(1387 November 2005)

2 Information about oil was recorded each year for 12 years.
The table shows the amount of oil produced (in billions of barrels) and the average price of oil (in £ per barrel).

Amount of oil produced (billions of barrels)	Average price of oil (£ per barrel)
7.0	34
11.4	13
10.8	19
11.3	12
9.6	23
8.2	33
7.7	30
10.9	12.5
8.0	28.5
9.9	13.5
9.2	26.5
9.4	15.5

a Draw a scatter graph to show the information in the table.

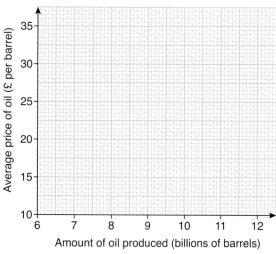

Amount of oil produced (billions of barrels)

b Describe the correlation between the average price of oil and the amount of oil produced.

c Draw a line of best fit on the scatter graph.

In another year the amount of oil produced was 10.4 million barrels.

d Use your line of best fit to estimate the average price of oil per barrel in that year.

(1384 June 1997)

3 The table shows the body temperature, in °C, and the pulse rate, in beats/min of 10 animals.

Body temperature (°C)	Pulse rate (beats/min)
36.5	30
38.6	50
37.5	160
36.7	110
39.7	80
38	40
39	200
38	80
38.9	100
37	70

a Draw a scatter graph to show the information in the table.

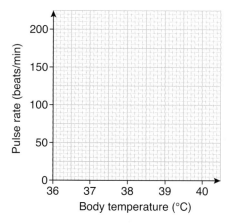

Body temperature (°C)

b Which of these terms best describes the relationship between body temperature and pulse rate?

positive correlation
negative correlation   zero correlation

Chapter 4 Fractions

Exercise 4A

1 Copy the fraction and fill in the missing number to make the fractions equivalent.

a $\frac{1}{6} = \frac{}{12}$

b $\frac{2}{3} = \frac{}{12}$

c $\frac{2}{5} = \frac{}{15}$

d $\frac{4}{5} = \frac{}{25}$

e $\frac{9}{10} = \frac{36}{}$

2 Write each fraction in its simplest form.

a $\frac{3}{12}$ **b** $\frac{12}{15}$ **c** $\frac{8}{10}$

d $\frac{48}{60}$ **e** $\frac{90}{150}$

3 Write each set of fractions in order. Start with the smallest fraction.

a $\frac{1}{4}, \frac{3}{8}$ **b** $\frac{5}{6}, \frac{2}{3}$

c $\frac{11}{30}, \frac{2}{5}, \frac{4}{15}$ **d** $\frac{1}{4}, \frac{3}{20}, \frac{2}{5}$

e $\frac{5}{8}, \frac{3}{4}, \frac{7}{16}$ **f** $\frac{9}{32}, \frac{3}{4}, \frac{7}{16}, \frac{5}{8}$

g $\frac{11}{40}, \frac{1}{2}, \frac{9}{20}, \frac{3}{5}$ **h** $\frac{3}{4}, \frac{5}{6}, \frac{7}{12}, \frac{13}{24}$

4 Ed and Jo have identical cans of cola. Ed drinks $\frac{4}{5}$ of his cola. Jo drinks $\frac{7}{9}$ of her cola. Who has drunk the most? You must give a reason for your answer.

5 Change these mixed numbers to improper fractions.

a $1\frac{4}{5}$ **b** $2\frac{2}{7}$ **c** $1\frac{3}{8}$

d $6\frac{2}{9}$ **e** $3\frac{11}{12}$

6 Change these improper fractions to mixed numbers.

a $\frac{7}{5}$ **b** $\frac{8}{3}$ **c** $\frac{13}{6}$

d $\frac{29}{3}$ **e** $\frac{89}{10}$

7 Write these improper fractions as mixed numbers in their simplest form.

a $\frac{34}{4}$ **b** $\frac{20}{6}$ **c** $\frac{34}{10}$

d $\frac{38}{8}$ **e** $\frac{52}{12}$

8 Write these fractions in order of size. Start with the smallest fraction.

$\frac{9}{16}$ $\frac{3}{4}$ $\frac{1}{2}$ $\frac{5}{8}$

(1388 March 2006)

Exercise 4B

Give each answer as a mixed number or a fraction in its simplest form.

1 Work out

a $\frac{1}{5} + \frac{1}{2}$ **b** $\frac{1}{6} + \frac{1}{4}$ **c** $\frac{1}{3} + \frac{2}{5}$

d $\frac{3}{7} + \frac{1}{5}$ **e** $\frac{4}{5} + \frac{1}{10}$ **f** $\frac{1}{8} + \frac{5}{6}$

2 Work out

a $\frac{3}{8} + \frac{2}{3}$ **b** $\frac{3}{5} + \frac{2}{3}$ **c** $\frac{3}{7} + \frac{2}{5}$

d $\frac{7}{9} + \frac{2}{3}$ **e** $\frac{7}{8} + \frac{5}{12}$ **f** $\frac{8}{15} + \frac{3}{10}$

3 Work out

a $\frac{1}{3} - \frac{1}{5}$ **b** $\frac{1}{2} - \frac{1}{8}$ **c** $\frac{1}{4} - \frac{1}{20}$

d $\frac{2}{5} - \frac{1}{4}$ **e** $\frac{7}{8} - \frac{1}{4}$ **f** $\frac{4}{5} - \frac{1}{3}$

4 Work out

a $\frac{2}{3} - \frac{3}{5}$ **b** $\frac{7}{9} - \frac{2}{3}$ **c** $\frac{13}{15} - \frac{3}{5}$

d $\frac{9}{10} - \frac{3}{4}$ **e** $\frac{11}{12} - \frac{5}{8}$ **f** $\frac{7}{9} - \frac{2}{5}$

5 Work out

a $\frac{4}{7} + \frac{2}{3}$ **b** $\frac{8}{9} - \frac{3}{5}$

c $\frac{7}{8} + \frac{5}{6}$ **d** $\frac{4}{5} - \frac{1}{3} + \frac{1}{2}$

e $\frac{11}{12} + \frac{3}{8} - \frac{5}{6}$ **f** $\frac{11}{18} + \frac{1}{6} - \frac{3}{4}$

Exercise 4C

 Give each answer as a mixed number or a fraction in its simplest form.

1 Work out

a $3\frac{1}{8} + 2\frac{5}{8}$ **b** $4\frac{2}{5} + 2\frac{1}{2}$ **c** $3\frac{1}{6} + \frac{7}{12}$

d $5\frac{5}{8} + 2\frac{3}{4}$ **e** $7\frac{2}{3} + 3\frac{7}{9}$ **f** $4\frac{5}{6} + 2\frac{2}{3}$

2 Work out

a $3\frac{8}{9} - 1\frac{2}{3}$ **b** $5\frac{4}{5} - 2\frac{1}{4}$

c $6\frac{5}{8} - 4\frac{2}{5}$ **d** $4\frac{9}{10} - 3\frac{1}{2}$

3 Work out

a $4 - \frac{5}{7}$ **b** $8 - \frac{5}{9}$

c $7 - 3\frac{3}{5}$ **d** $10 - 6\frac{5}{12}$

4 Work out

a $4\frac{1}{2} - 1\frac{3}{4}$ **b** $5\frac{1}{6} - 3\frac{2}{3}$ **c** $5\frac{2}{7} - 2\frac{3}{5}$

d $4\frac{3}{8} - 2\frac{5}{6}$ **e** $6\frac{5}{8} - 2\frac{2}{3}$ **f** $8\frac{3}{5} - 5\frac{4}{9}$

5 Work out

a $6\frac{1}{2} + 2\frac{3}{8}$ **b** $7\frac{4}{5} - 2\frac{9}{10}$ **c** $4\frac{5}{8} + 2\frac{3}{4}$

d $15\frac{8}{15} - 8\frac{7}{10}$ **e** $7\frac{8}{9} + 3\frac{2}{3}$ **f** $5\frac{3}{10} - 2\frac{7}{8}$

Exercise 4D

 Give each answer in its simplest form.

1 Work out

  a $3 \times \frac{1}{5}$ **b** $4 \times \frac{1}{6}$ **c** $\frac{2}{3} \times 5$

  d $\frac{3}{8} \times 10$ **e** $6 \times \frac{5}{12}$ **f** $\frac{9}{20} \times 25$

2 Work out

  a $\frac{2}{3} \times \frac{1}{7}$ **b** $\frac{1}{5} \times \frac{2}{7}$ **c** $\frac{5}{9} \times \frac{3}{8}$

  d $\frac{8}{11} \times \frac{2}{5}$ **e** $\frac{7}{8} \times \frac{3}{4}$ **f** $\frac{6}{11} \times \frac{3}{7}$

3 Work out

  a $\frac{7}{8} \times \frac{4}{5}$ **b** $\frac{6}{7} \times \frac{3}{4}$ **c** $\frac{5}{9} \times \frac{3}{7}$

  d $\frac{4}{5} \times \frac{10}{11}$ **e** $\frac{32}{35} \times \frac{15}{16}$ **f** $\frac{63}{77} \times \frac{14}{27}$

4 Work out

  a $2\frac{1}{4} \times \frac{1}{3}$ **b** $3\frac{1}{2} \times \frac{3}{14}$ **c** $1\frac{2}{3} \times \frac{7}{10}$

  d $4\frac{1}{2} \times 1\frac{1}{3}$ **e** $4\frac{2}{5} \times 2\frac{1}{2}$ **f** $4\frac{2}{7} \times 1\frac{1}{5}$

5 Work out

  a $4\frac{2}{3} \times 2\frac{1}{4}$ **b** $3\frac{3}{5} \times 2\frac{1}{12}$ **c** $7\frac{1}{5} \times 1\frac{7}{8}$

  d $1\frac{1}{14} \times 1\frac{1}{20}$ **e** $6\frac{2}{3} \times 3\frac{4}{5}$ **f** $3\frac{5}{9} \times 1\frac{7}{8}$

6 Work out

  a $\frac{4}{5} \times \frac{3}{8} + \frac{1}{2}$ **b** $\frac{5}{7} - \frac{2}{3} \times \frac{9}{10}$

  c $(3\frac{1}{2} - 2\frac{2}{3}) \times 2\frac{2}{5}$ **d** $(3\frac{3}{8} + 2\frac{1}{4}) \times 1\frac{1}{3}$

7 Work out $60 \times \frac{2}{3}$

                                  (1387 June 2005)

8 a Work out $\frac{1}{4} \times \frac{2}{3}$
      Give your answer as a fraction in its simplest form.

  b Work out $2\frac{3}{8} - 1\frac{1}{4}$

                (1388 January 2002)

Exercise 4E

 Give each answer in its simplest form.

1 Work out

  a $\frac{4}{9} \div 2$ **b** $\frac{3}{5} \div 2$ **c** $\frac{6}{7} \div 3$

  d $\frac{3}{4} \div \frac{1}{2}$ **e** $\frac{7}{8} \div \frac{3}{4}$ **f** $\frac{9}{10} \div \frac{3}{5}$

2 Work out

  a $\frac{4}{5} \div \frac{5}{6}$ **b** $\frac{8}{9} \div \frac{2}{3}$ **c** $\frac{14}{25} \div \frac{2}{5}$

  d $\frac{21}{32} \div \frac{3}{8}$ **e** $\frac{42}{55} \div \frac{7}{22}$ **f** $\frac{18}{35} \div \frac{12}{49}$

3 Work out

  a $2\frac{2}{3} \div 4$ **b** $3\frac{3}{8} \div \frac{3}{4}$ **c** $7\frac{1}{2} \div 10$

  d $1\frac{7}{9} \div \frac{4}{5}$ **e** $4\frac{2}{5} \div \frac{11}{20}$ **f** $2\frac{5}{8} \div \frac{35}{36}$

4 Work out

  a $3\frac{1}{2} \div 1\frac{1}{4}$ **b** $1\frac{3}{7} \div 2\frac{1}{2}$ **c** $4\frac{4}{5} \div 1\frac{5}{7}$

  d $5\frac{1}{4} \div 4\frac{3}{8}$ **e** $7\frac{3}{9} \div 4\frac{2}{3}$ **f** $5\frac{5}{6} \div 10\frac{1}{2}$

5 Work out

  a $(\frac{5}{6} + \frac{3}{8}) \div \frac{1}{3}$ **b** $7 - \frac{6}{7} \div \frac{3}{5}$

  c $5\frac{1}{4} \div 3\frac{1}{2} \times 1\frac{2}{7}$ **d** $4\frac{1}{6} \div (3\frac{2}{3} - 2\frac{5}{9})$

6 Work out

  a $1\frac{7}{8} \times 5\frac{1}{3}$ **b** Work out $3\frac{1}{2} \div 2\frac{4}{5}$

                                    (1388 June 2003)

Exercise 4F

 1 Find

  a $\frac{1}{3}$ of 15   **b** $\frac{1}{6}$ of 24   **c** $\frac{1}{8}$ of 48

  d $\frac{1}{5}$ of 45   **e** $\frac{1}{4}$ of 120   **f** $\frac{1}{10}$ of 350

2 Find

  a $\frac{3}{5}$ of 30   **b** $\frac{2}{7}$ of 28   **c** $\frac{5}{6}$ of 12

  d $\frac{6}{7}$ of 28   **e** $\frac{4}{9}$ of 63   **f** $\frac{5}{8}$ of 56

3 Find

  a $\frac{3}{8}$ of £40   **b** $\frac{6}{7}$ of 140 cm

  c $\frac{3}{4}$ of 1500 m   **d** $\frac{2}{3}$ of 135 g

  e $\frac{2}{7}$ of 175 km   **f** $\frac{2}{9}$ of £5238

4 Find

  a $\frac{2}{3}$ of 4   **b** $\frac{3}{7}$ of 5   **c** $\frac{4}{5}$ of 2

  d $\frac{7}{12}$ of 20   **e** $\frac{9}{16}$ of 24   **f** $\frac{3}{20}$ of 50

5 Find

  a $\frac{1}{4}$ of $\frac{2}{5}$ **b** $\frac{3}{7}$ of $\frac{5}{6}$ **c** $\frac{2}{9}$ of $\frac{3}{8}$

  d $\frac{9}{11}$ of $\frac{5}{18}$ **e** $\frac{5}{6}$ of $\frac{18}{25}$ **f** $\frac{4}{9}$ of $\frac{3}{14}$

6 Find

  a $\frac{2}{3}$ of £582   **b** $\frac{4}{5}$ of 785 kg

  c $\frac{2}{7}$ of 294 cm   **d** $\frac{13}{15}$ of 540 g

  e $\frac{23}{25}$ of 1775 m   **f** $\frac{67}{100}$ of £70

7 Find

  a $\frac{3}{4}$ of £13   **b** $\frac{2}{5}$ of 18 m

  c $\frac{7}{12}$ of 168 cm   **d** $\frac{7}{10}$ of 295 g

  e $\frac{5}{8}$ of £84   **f** $\frac{9}{16}$ of 10 km

Exercise 4G

1 Poppy travels $1\frac{3}{4}$ miles to the station. She then travels a further $3\frac{1}{5}$ miles by train. How far has she travelled altogether?

2 There are 1200 students in a school. $\frac{5}{8}$ of the students are girls. Work out the number of boys in the school.

3 Charlie records three television programmes. The first programme is $1\frac{1}{2}$ hours long, the second programme is $1\frac{1}{3}$ hours long and the third programme lasts for $\frac{3}{4}$ of an hour. He uses a 4 hour video. How much of the video will not be used?

4 Lizzie buys 5 m of material. She needs $\frac{4}{5}$ of a metre of material to make one cushion. Work out how many cushions she can make from the 5 m of material.

5 Alex has 120 books. $\frac{1}{3}$ of her books are romances, $\frac{1}{4}$ of her books are thrillers and the rest are science fiction. How many science fiction books does Alex have?

6 A rectangle is $3\frac{4}{5}$ m long and $2\frac{3}{8}$ m wide. Work out the area of the rectangle.

7 A floor is covered with tiles. $\frac{3}{5}$ of the tiles are blue, $\frac{1}{3}$ of the tiles are black and the rest of the tiles are white. What fraction of the tiles are white?

8 Last season, Excel United won $\frac{3}{10}$ of its matches, lost $\frac{1}{4}$ and drew the rest. What fraction of its matches did it draw?

9 $\frac{7}{8}$ of the yoghurts in a pack are flavoured yoghurts. $\frac{1}{3}$ of the flavoured yoghurts are apricot yoghurts. What fraction of the yoghurts are apricot flavoured?

10 Miriam gave her mother **two** £5 notes. Miriam said, 'This is $\frac{1}{4}$ of my day's pay.' Work out Miriam's pay that day.

          (1388 January 2004)

11 Some students each chose one PE activity. $\frac{1}{5}$ of the students chose swimming. $\frac{3}{8}$ of the students chose tennis. All the rest chose cricket. What fraction of the students chose cricket?

          (1387 November 2005)

Chapter 5 Expressions and sequences

Exercise 5A

1 There are b black socks and w white socks in a drawer. Write down an expression, in terms of b and w for the total number of socks in the drawer.

2 There are m children in a room. n children are seated and the rest are standing. Write down an expression, in terms of m and n, for the number of children standing.

3 There are 5 biscuits in each packet. Maria buys y packets of biscuits and then eats b biscuits. Write down an expression, in terms of y and b, for the total number of biscuits Maria has left.

4 In football matches 3 points are awarded for a win, 1 point is awarded for a draw and no points are awarded for a loss. Jim's football team will play 20 matches next year. He expects his team to win p of them, draw q of them and lose the rest of the matches. Write down an expression, in terms of p and q, for

  a the total number of matches Jim expects his team lose next year.

  b the total number of points Jim expects his team to be awarded next year.

5 Work out the value of each of these expressions, when $x = 3$, $y = -5$ and $z = -2$

a $3x + y$
b $x + 4z$
c $y - 3z$
d $2x - 2y$
e $2x + 3z$
f $x - y + z$
g $4x + 3y - 2z$
h $3y - 8z - 5x$
i $3z - 5y + 4x$
j $8x - 2y - z$

6 Simplify

a $5x + 2x - 3y + y$
b $8x - x - 4y - 5y$
c $3x - 4y - 5x + 7y$
d $2x + x + 2y - 6x - 6y - 3y$
e $4x - 5y - 2y - 3x$
f $x - 5y - 7x + 3y + 6x$
g $2a + 3b - 4a - 5a$
h $7m - n - 6n + m$
i $3e + f - 6e + 5f + 2e$
j $4c + 3d - 4d + c + d - 5c$
k $7e - 2f - 3 - 6f - e - 4$
l $3 - s - 2t - 6t + 8s - 4$

7 Find the value of the number p and the value of number q so that the expression $6x - 2y - 5y - 3x + px + qy$ simplifies to 0.

Exercise 5B

1 Write these expressions in their simplest form

a $a \times b \times c$
b $p \times 5 \times r$
c $m \times 3m$
d $4s \times t \times u$
e $x \times x \times 4x$
f $y \times 2y \times y^2$
g $p \times p \times 2 \times q \times q$

2 Simplify

a $4 \times 3a$
b $5p \times 3p$
c $8g \times h^2$
d $y \times 5z$
e $6s \times 7t$
f $8m \times 3n$
g $5c \times 2 \times 7d$
h $3d \times 4e \times 5$
i $x \times 3x$
j $c \times d \times 4e$
k $2u \times 5v \times w$
l $a \times 3a \times 6a$

m $5a \times 2b \times a$
n $2xy \times 4y$
o $a \times b + a \times b$
p $y \times y + y \times y$
q $p \times p + p \times p + p \times p$

3 Work out the value of each of these expressions, when $x = -2$

a $x^2 + 3$
b $x^3 - 4x$
c $x^4 + 2x^2$
d $x^5 - 3x$
e $x^6 + 2x^3$

4 Work out the value of these expressions when $a = -2$.

a $\dfrac{6a^4}{a^2}$
b $5a^2 + 8a - 7$
c $\dfrac{6a^2 + 8a^3}{a^2}$
d $\dfrac{5a^3 + 4a}{3a}$

5 Work out the value of each of these expressions when $a = 2$, $b = 4$, $c = -2$

a $3a + bc$
b $2b^2 - 7$
c abc
d $3ab - c$
e $6bc - 3ac$
f $\dfrac{a + b}{c}$
g $\dfrac{b + 2c^2}{a^3}$

6 Work out the value of each of these expressions, when $p = -3$, $q = -4$, $r = 6$

a pq^2
b $2rp^2 - 5q^2$
c $2p^3 - q^3$
d $\dfrac{p^4 - 2r^2}{pq}$
e $11 - \dfrac{pq^2r}{p + qr}$

7 Work out the value of the expression $4(x + 2)^3$ when $x = -5$.

Exercise 5C

1 Simplify

a $x^2 \times x^6$
b $y^8 \times y$
c $3z \times 4z^5$
d $2a^5 \times 4a^{-2}$
e $x^7 \div x^4$
f $y^9 \div y$
g $a^6 \div a^5$
h $z^5 \div z^{-1}$

i $8q^5 \div 2q^{-2}$ **j** $2y \times y^2 \times 3y^6$

k $q^4 \times q \div q^3$ **l** $5q \times 6q^3 \div 2q^{-2}$

m $(4 \times x \times x^3) + (x^6 \div x^2)$

n $(24y^6 \div 2y^2) - (3y^3 \times 2y)$

o $(4x^3 \times 4x^3) + (8x^{10} \div 2x^4) - (2x^2)^3$

2 Simplify

a $a^3b^4 \times ab^3$ **b** $3c^2d^4 \times 5c^6d$

c $4e^2g^4 \times 3g^6e^8$ **d** $15x^8y^4 \div 3x^4y^{-2}$

e $12x^3y^6 \div 2y^3x$ **f** $24r^4s^6 \div 6s^{-3}r^2$

3 Find the value of

a $3x^0 + 2$ **b** $(2xy)^0$

c $(x + y + 1)^0$ **d** $x^0 + y^0 + 1^0$

4 Write as a power of y

a $\dfrac{1}{y^8}$ **b** $\dfrac{1}{y^8 \div y^2}$

c $\dfrac{1}{y^2 \times y}$ **d** $\dfrac{1}{y \div y^7}$

5 a Simplify $x^4 \div x^9$

b Simplify $3w^5y^2 \div 4w^3y^4$

(1388 March 2006)

6 Simplify

a $(x^4)^2$ **b** $(3y^3)^3$

c $(x^5y^2)^3$ **d** $(2pq^2)^4$

e $(x^3)^{-3}$ **f** $(2y^{-3})^{-2}$

g $(z^{-5})^{-4} \times z^3$ **h** $(-2x^{-4})^{-3} \div x^8$

i $(x^4)^0 + x^2$ **j** $y(y^0)^3$

k $(3a^2b^{-4})^{-3} \times (3b^{-3}a)^3$

7 Simplify

a $(3x^2y)^3$ **b** $(2t^{-3})^{-2}$

(1387 November 2005)

Exercise 5D

1 An expression for the nth term of the arithmetic sequence 5, 7, 9, 11, ... is $2n + 3$

a Find the 10th term of the sequence.

b Find the 100th term of the sequence.

2 Here are the first five terms of an arithmetic sequence. 1, 3, 5, 7, 9

a Write down, in terms of n, an expression for the nth term of this sequence.

b 1 is the first odd number. Find the 100th odd number.

3 The first five terms of an arithmetic sequence are 9, 13, 17, 21, 25. Find, in terms of n, an expression for the nth term of this sequence.

(1388 March 2006)

4 The first five terms of an arithmetic sequence are 2, 9, 16, 23, 30. Find, in terms of n, an expression for the nth term of this sequence.

(1388 January 2004)

5 The first four terms of an arithmetic sequence are 71, 74, 77, 80. Find, in terms of n, an expression for the nth term of this sequence.

(1388 January 2002)

6 Here are the first five terms of an arithmetic sequence. 40, 20, 0, −20, −40

a Write down, in terms of n, an expression for the nth term of this sequence.

b Find the 25th term of the sequence.

7 a Write down, in terms of n, an expression for the nth term of the arithmetic sequence 100, 98, 96, 94, ...

b Write down, in terms of n, an expression for the nth term of the arithmetic sequence 0, −6, −12, −18, ...

c Use your answers to parts **a** and **b** to find, in terms of n, an expression for the nth term of the arithmetic sequence 100, 92, 84, 76, ...

8 Jason writes down the arithmetic sequence
$-26, -22, -18, -14, \ldots$
Karen writes down the arithmetic sequence
$972, 974, 976, 978, \ldots$

  a Write down, in terms of n, an expression for the nth term of the arithmetic sequence $-26, -22, -18, -14, \ldots$

  b Find the difference between the 1000th term of Jason's arithmetic sequence and the 1000th term of Karen's arithmetic sequence.

  c The kth term of Jason's arithmetic sequence is the same number as the kth term of Karen's arithmetic sequence. Find the value of k.

Chapter 6 Measure

Exercise 6A

1 Rory uses a scale of $1:50$ to draw a plan of a house to scale.

  a On the plan, the kitchen is 10 cm long and 7.6 cm wide.
Find, in metres, the real length and width of the kitchen.

  b The dining room is 7 m long and 5.8 m wide.
Find, in centimetres, the length and width of the dining room on the plan.

2 Darren uses a scale of $1:48$ to make a model of a train locomotive.

  a The length of the model is 41 cm.
Find, in metres, the real length of the locomotive.

  b The diameter of the wheels of the locomotive is 1.2 m
Find, in centimetres, the diameter of the wheels on the model.

3 A map is drawn to a scale of $1:20\,000$

  a On the map, the distance between Preety's home and her school is 15 cm.
Work out the real distance, in kilometres, between her home and her school.

  b The real distance between two schools is 4.6 km.
Work out, in centimetres, their distance apart on the map.

4 A map is drawn to a scale of $1:200\,000$

  a On the map, the distance between Canterbury and Dover is 13 cm.
Work out the real distance in kilometres, between Canterbury and Dover.

  b The real distance between Sheffield and Doncaster is 35 km.
Work out, in centimetres, their distance apart on the map.

5 Simone made a scale model of a "hot rod" car on a scale of $1:12.5$
The height of the model car is 10 cm.

  a Work out the height of the real car.

The length of the real car is 5 m.

  b Work out the length of the model car. Give your answer in centimetres.

The angle the windscreen made with the bonnet of the real car is $140°$.

  c What is the angle the windscreen makes with the bonnet of the model car?
(1385 June 1998)

6 Natasha makes a scale drawing of a field.
The real length of the field is 300 m.
On the drawing, the length of the field is 15 cm.
Find, as a ratio, the scale of her drawing.

7 The distance between two towns is 9 km.
On a map, the distance between the two towns is 18 cm.
Find, as a ratio, the scale of the map.

Exercise 6B

1 Sunil drove 70 miles in $1\frac{1}{4}$ hours. Work out his average speed in mph.

2 Suhail cycles 117 km in 4 hours 30 minutes. Work out his average speed in km/h.
(4400 May 2004)

3 Fiona drove for $3\frac{3}{4}$ hours at an average speed of 60 km/h.
Work out the distance she travelled.

4 Dale cycled 156 km at an average speed of 24 km/h.
Work out the time he took.

5 Olivia walked for 3 hours 20 minutes at an average speed of 6 km/h.
Work out the distance she walked.

6 Mary drove from her home to her friend's house.
The distance she drove was $31\frac{1}{2}$ km.
She left home at 10 40 am and arrived at her friend's house at 11 15 am.
Work out Mary's average speed for her journey. Give your answer in km/h.

7 Sally drove the 85 miles from Birmingham to Bristol at an average speed of 50 mph.
Work out, in hours and minutes, how long it took her.

8 Niall went on a cycle ride.
For the first part of the ride, he cycled 40 km at an average speed of 16 km/h.
For the rest of his ride, he cycled for 2 hours at an average speed of 12 km/h.

  a Work out the total distance Niall cycled.

  b Work out the total time taken by Niall on his ride.

  c Work out Niall's average speed for the whole ride.
Give your answer in km/h, correct to one decimal place.

9 In the 2004 Olympic Games, the winner's time in the women's 400 m race was 49.4 s.
Work out her average speed in

  a m/s   **b** km/h.

Give your answers correct to 1 decimal place.

Exercise 6C

1 A piece of silver has a mass of 52.5 grams and a volume of 5 cm^3.
Work out the density of the silver.

2 A cylinder has a volume of 2260 cm^3. The cylinder is made of material that has a density of 1.5 g/cm^3. Work out the mass of the cylinder.
(1388 April 2005)

3 Aluminium has a density of 2700 kg/m^3.
Work out the volume of a piece of aluminium with a mass of 8100 kg.

4 A block of wood has a volume of 140 cm^3.
The wood has a density of 1.2 grams per cm^3.
Work out the mass of the block of wood.
(1388 April 2005)

5 A copper rod has a mass of 427.2 grams and a volume of 48 cm^3.

  a Work out the density of copper.

Another copper rod has a mass of 1.78 kg.

  b Work out the volume of this copper rod.

6 An ice hockey puck is made out of rubber with a density of 1.5 grams per cm^3.
The volume of the puck is 113.4 cm^3.
Work out the mass of the ice hockey puck.
Give your answer correct to the nearest gram.

Exercise 6D

1 x, y and z represent lengths.
For each of these expressions, state whether it could represent a length, an area, a volume or none of these.
(Numbers have no dimensions.)

  a x^2y **b** $2y + 3z$ **c** $xy + z$

  d πyz **e** $x(y + z)$ **f** $\dfrac{xy}{z}$

2 x, y and z represent lengths.
Here are some expressions.

$$2xy + 3yz \qquad x^2(y + z) \qquad \frac{x^2y}{z}$$

$$x^2 + 5y \qquad \pi x + 4yz \qquad \frac{xyz}{6}$$

   a Write down the expressions which could represent an area.
   b Write down the expressions which could represent a volume.

3

$\dfrac{\pi r^2}{x}$	$\pi(r + x)$	$\pi r + r$	$\dfrac{\pi r^3}{x}$	$\pi r^2 + rx$	$\dfrac{r^2}{\pi x}$

Here are some expressions.
The letters r and x represent lengths. π is a number which has no dimension. Two of the expressions could represent areas.

   a Copy the table and tick the boxes (✓) underneath the two expressions which could represent areas.

Here are four more expressions.

πr^3	$\pi r^4 + \pi x$	$\dfrac{\pi r^4}{x}$	$\pi r^2 + \pi rx$

One of these four expressions cannot represent a length or an area or a volume.

   b Copy the table and put a cross in the box (✗) underneath the one expression which cannot represent a length or an area or a volume.
   c Rearrange the formula $y = r + 3x$ to make x the subject.
      (1385 Novemeber 2001)

Exercise 6E

1 Change 30 miles to kilometres.

2 Change 6 inches to centimetres.

3 Change 120 centimetres to feet.

4 Change 10 centimetres to inches.

5 Change 35 pints to litres.

6 Change 44 pounds to kilograms.

7 Change 180 litres to gallons.

8 Change 20 kilometres to miles

9 Change 5 kg to pounds.   *(1388 June 2004)*

10 A car is travelling at a speed of 60 miles per hour. Change 60 miles per hour to kilometres per hour.

11 An aeroplane flies 2400 km. Change 2400 kilometres to miles.

12 A baby weighs 3.5 kilograms. Change 3.5 kilograms to pounds.

13 A turkey weighs 16.5 pounds. Change 16.5 pounds to kilograms.

14 A book measures 9 inches. Change 9 inches to centimetres.

15 John buys 9 gallons of petrol. Change 9 gallons to litres.

Chapter 7 Decimals and fractions

Exercise 7A

1 Write each of the following sets of numbers in order of size. Start with the smallest number each time.
   a 0.85, 0.58, 0.5, 0.8
   b 59.12, 59.21, 59.2, 59.11
   c 7.263, 7.236, 7.632, 7.623, 7.362
   d 0.183, 0.831, 0.138, 0.18, 0.31
   e 4.062, 4.026, 4.002, 4.06, 4.022
   f 9.317, 9.71, 9.713, 9.137, 9.13

2 Work out
   a 8.3 + 6.5   **b** 3.65 + 2.84
   c 56.39 + 7.06   **d** 15.4 + 3.78
   e 8 + 2.74 + 0.9
   f 16.5 + 8.64 + 2.347

3 Work out

   a $9.48 - 3.16$ **b** $24.52 - 12.36$

   c $86.4 - 13.27$ **d** $17 - 6.8$

   e $65 - 31.46$ **f** $87.2 - 19.63$

4 Work out $7.6 - 4.83$

(1388 January 2005)

5 Work out

   a 7.24×10 **b** 82.546×100

   c 13.2×100 **d** 53.2×1000

   e 0.0031×10 **f** 0.087×1000

   g 0.9×100 **h** 0.00395×1000

6 Work out

   a 2.53×0.3 **b** 7.32×0.2

   c 5.63×0.4 **d** 8.04×0.6

   e 5.19×0.05 **f** 6.34×0.03

7 Work out

   a 5.3×2.1 **b** 7.2×1.3

   c 6.3×0.52 **d** 0.34×0.52

   e 9.5×4.3 **f** 0.76×5.3

8 The cost of a calculator is £6.79. Work out the cost of 28 of these calculators.

(1387 June 2005)

9 Nick takes 26 boxes out of his van.
The weight of each box is 32.9 kg.
Work out the **total** weight of the 26 boxes.

(1387 June 2004)

10 Work out

   a $34 \div 10$ **b** $45.6 \div 100$

   c $83 \div 100$ **d** $5.3 \div 1000$

   e $0.45 \div 10$ **f** $235.7 \div 100$

   g $36 \div 1000$ **h** $0.4 \div 10$

11 Work out

   a $46 \div 0.2$ **b** $8.4 \div 0.3$

   c $3.12 \div 0.04$ **d** $87.5 \div 0.05$

   e $8 \div 0.02$ **f** $0.085 \div 0.05$

   g $14.31 \div 0.9$ **h** $1.498 \div 0.07$

12 Three shirts cost £25.80. Work out the price of one shirt.

13 Five drinks cost £6.75. Work out the price of one drink.

14 Six people share a lottery win of £1443.12 equally. Work out how much money each person will get.

15 A 5 kg bag of peanuts is divided up into smaller bags each containing 0.2 kg of peanuts. Work out the number of smaller bags that can be filled.

Exercise 7B

1 Given that $5.7 \times 3.4 = 19.38$ work out

   a 57×34 **b** 570×3.4

   c 0.57×0.34 **d** 5700×34

2 Give that $92.3 \div 6.5 = 14.2$ work out

   a $923 \div 6.5$ **b** $92\,300 \div 6.5$

   c $0.923 \div 6.5$ **d** $923 \div 0.065$

3 Given that $8.2 \times 17.5 = 143.5$ work out

   a 82×17.5 **b** 0.82×1.75

   c 82×0.175 **d** 0.082×0.175

4 Given that $95.76 \div 6.3 = 15.2$ work out

   a $0.9576 \div 6.3$ **b** $95.76 \div 6300$

   c $957.6 \div 0.063$ **d** $9.576 \div 630$

5 Given that $\dfrac{13.2 \times 5.7}{3.8} = 19.8$ work out

   a $\dfrac{132 \times 57}{3.8}$ **b** $\dfrac{1.32 \times 0.057}{3.8}$

   c $\dfrac{1.32 \times 5.7}{38}$ **d** $\dfrac{1.32 \times 570}{0.38}$

6 Given that $\dfrac{16.4 \times 7.2}{4.8} = 24.6$ work out

   a $\dfrac{1640 \times 72}{4.8}$ **b** $\dfrac{16.4 \times 7.2}{480}$

   c $\dfrac{164 \times 0.72}{480}$ **d** $\dfrac{0.164 \times 0.72}{480}$

7 Given that $54 \times 17 = 918$ work out

   a 0.54×170 **b** 540×0.17

   c $918 \div 0.17$ **d** $9.18 \div 170$

8 Given that $2158 \div 83 = 26$ work out

   a $21.58 \div 8.3$ **b** $21\,580 \div 0.83$

   c 0.26×83 **d** $21.58 \div 260$

9 $32 \times 129 = 4128$

   Write down the value of

   i 3.2×1.29 **ii** $0.32 \times 129\,000$

                                      (1388 January 2003)

10 Using the information that

   $38 \times 323 = 12\,274$

   find the value of

   i 0.38×323 **ii** $12\,274 \div 380$

   iii 37×323

                                     (1388 March 2004)

11 Using the information that

   $73 \times 154 = 11\,242$

   write down the value of

   i 7.3×1.54 **ii** $112\,420 \div 0.73$

                           (1388 January 2005)

12 Using the information that

   $65 \times 423 = 27\,495$ find the value of

   i 0.65×4230 **ii** $274.95 \div 650$

                           (1388 March 2006)

Exercise 7C

1 Write each of the decimals as a fraction in its simplest form.

   a 0.4 **b** 0.32

   c 0.239 **d** 0.06

   e 0.008 **f** 0.125

   g 43.2 **h** 10.003

2 Write the following fractions as decimals.

   a $\frac{3}{10}$ **b** $\frac{13}{100}$ **c** $\frac{9}{100}$

   d $\frac{129}{1000}$ **e** $\frac{7}{1000}$

3 Write the following as equivalent fractions and then as decimals.

   a $\frac{3}{5} = \frac{}{10}$ **b** $\frac{9}{20} = \frac{}{100}$ **c** $\frac{21}{50} = \frac{}{100}$

   d $\frac{13}{500} = \frac{}{1000}$ **e** $\frac{8}{25} = \frac{}{100}$

4 Write down the following fractions as decimals.

   a $\frac{1}{4}$ **b** $\frac{1}{10}$ **c** $\frac{1}{100}$

   d $\frac{3}{4}$ **e** $\frac{1}{2}$

5 Use short division to change these fractions to decimals.

   a $\frac{4}{5}$ **b** $\frac{1}{8}$

6 Use a calculator to change these fractions to decimals.

   a $\frac{1}{16}$ **b** $\frac{17}{40}$ **c** $\frac{123}{160}$

   d $\frac{19}{25}$ **e** $\frac{21}{32}$

7 Use a calculator to change these fractions to decimals.

   a $\frac{5}{6}$ **b** $\frac{8}{11}$ **c** $\frac{7}{15}$

   d $\frac{23}{24}$ **e** $\frac{3}{7}$

8 $\frac{1}{3}$ $\frac{2}{5}$ $\frac{5}{8}$ $\frac{6}{10}$ $\frac{7}{12}$ $\frac{9}{15}$

   Maria converted each of these fractions to decimals.

   Put a ring around each fraction which gave a recurring decimal.

                       (1388 November 2005)

Exercise 7D

In questions **1** to **8** convert each of the recurring decimals to a fraction. Give each fraction in its simplest form.

1 $0.55555\ldots$ **5** $0.8\dot{3}$

2 $0.363636\ldots$ **6** $0.3\dot{1}\dot{2}$

3 $0.187187\ldots$ **7** $5.7\dot{9}$

4 $0.\dot{6}\dot{5}$ **8** $3.4\dot{5}\dot{2}$

9 Convert the recurring decimal $0.\dot{0}1\dot{3}$ to a fraction.   *(1388 January 2004)*

10 Express $0.3\dot{2}\dot{8}$ as a fraction in its simplest form.   *(1388 March 2006)*

11 Given that $\frac{7}{11} = 0.\dot{6}\dot{3}$ write the recurring decimal $0.2\dot{6}\dot{3}$ as a fraction.

12 Given that $\frac{8}{33} = 0.\dot{2}\dot{4}$ write the recurring decimal $0.5\dot{3}2\dot{4}$ as a fraction.

Exercise 7E

1 Round these numbers to one significant figure

  a 563   **b** 8924   **c** 611

  d 24   **e** 0.0546   **f** 0.00319

2 Round these numbers to three significant figures

  a 5613   **b** 34.186   **c** 0.98134

  d 35.468   **e** 3.1709   **f** 0.0091456

3 Round these to the number of significant figures given in the brackets

  a 923 (1)   **b** 67.354 (3)

  c 128 (2)   **d** 0.0345 (1)

  e 0.03762 (2)   **f** 839 524 (3)

  g 0.00263 (1)   **h** 0.85625 (3)

  i 297 600 (2)   **j** 3.247 (3)

  k 43.73 (2)   **l** 7356 (1)

4 Use your calculator to work out the value of the following. Give each answer correct to 3 significant figures.

  a $6782 \div 43$ **b** 561×29

  c 0.034×0.457 **d** $\dfrac{45.1 \times 63.6}{0.09}$

  e $\dfrac{45.21}{67.2 - 7.93}$ **f** $\dfrac{73.51 + 26.3}{5.34 - 2.9}$

5 a Write the number 56 392 correct to one significant figure.

  b Write the number 0.0436 correct to one significant figure.   *(1388 January 2005)*

6 a Write the number 7623 correct to 1 significant figure.

  b Write the number 0.00821 correct to 2 significant figures.

  c Use your calculator to work out

    $\dfrac{15.1 + 4.82}{6.2 - 3.7}$

    Write down all the figures on your calculator display.

    (1388 November 2005)

Chapter 8 Expanding brackets and factorising

Exercise 8A

1 Expand

  a $4(x + 1)$ **b** $2(x - 1)$

  c $5(1 + k)$ **d** $3(x + y)$

  e $3(2 - q)$ **f** $4(s - t)$

  g $5(m - 7)$ **h** $2(p - q)$

  i $3(2x + 3)$ **j** $2(5y - 2)$

  k $4(3 + 4z)$ **l** $5(1 - 3p)$

  m $2(3q + 7r)$ **n** $3(3s - t)$

  o $5(5u - 4v)$ **p** $4(p + 2 + 3q)$

  q $6(1 + 2e + f)$ **r** $5(3r + 2s - 4)$

  s $4(2p - 3q - 4r)$

2 Expand

  a $x(x + 2)$ **b** $y(y - 3)$

  c $2a(a + 1)$ **d** $b(6 - 5b)$

  e $c(7c + c^2)$ **f** $d(2e - d)$

  g $-3(x + 2)$ **h** $-y(x + 4)$

  i $-4(3y^2 - 2y)$ **j** $-2(a^2 - 3a - 4)$

  k $3x(2x^2 - 4x - 5)$ **l** $-x(x - 4)$

3 Expand and simplify $5(2x + 3) - 2(x - 1)$

    (1385 May 2002)

4 Expand and simplify $2(3x + 4) - 3(4x - 5)$

    (1387 June 2003)

5 Expand and simplify

  a $3(z + 2) + 1$

  b $5(y + 3) - 3y$

  c $4(x + 1) - 7$

  d $2(3 + w) - 7w$

  e $4(u + v) + 1(u + 3v)$

  f $4(2a + 3b) + 3(4a + 3b)$

  g $2(4c + 3d) + 4(2c - 3d)$

  h $2(2e + 3f) + 3(2f - e)$

  i $6(4g - 3h) + 5(3h - 2g)$

6 Expand and simplify

a $6x - 5(x + 2)$

b $7y - 4(y - 3)$

c $9a - 2(3a + 4)$

d $4b - 1 - 6(b - 2)$

e $x(x + 3) + 2(x + 1)$

f $y(y + 4) - 2(y + 1)$

g $z(z - 4) - (z - 3)$

h $k(k + 5) - 6(k - 2)$

i $-2x(x - 4y) + y(2x + 3y)$

Exercise 8B

1 Factorise these expressions.

a $5x + 10$

b $9y + 3$

c $35z - 7$

d $33a + 11b$

e $6c + 8d$

f $9e - 15f$

g $12g - 18h$

h $16m + 24n$

i $10p - 15q$

j $20r + 24s$

k $8t - 12u + 20$

l $4v - 8w - 10$

m $36x - 30y - 42z$

n $ap + aq$

o $2cd + ed$

p $3ab - bc$

q $p - 2qp$

r $2cd - 3de + 4d$

s $y^2 - 8y$

t $x^2 + x$

u $k - k^2$

v $3y + y^2$

w $2y^2 + 3y$

x $7y - 4y^2$

y $x^3 - 3x$

z $t^2 + pt - 8t$

2 Factorise completely

a $3abc + 3ad$

b $4pr + 2qr + 6rs$

c $4x^2 - 8xy^2$

d $5ac - cba$

e $8pr + 6p^2r$

f $ac - 2ac^2$

g $y^4 - 2y^2$

h $9d^2 + 6d^3e$

i $abc^3 + a^2b$

j $16xy^2 - 24y$

k $4xy^2 - 5x^2y^3$

l $10p^2q - 15qp^3$

m $8c^2d + 16dc^3 + 12cd^2$

n $(2p^2q)^2 + 8p^3q^2$

Exercise 8C

1 Expand and simplify $(y + 3)(y + 4)$

(1388 March 2006)

2 Expand and simplify

a $(x + 4)(x + 1)$

b $(y + 3)(y + 7)$

c $(z + 2)(z - 1)$

d $(x + 5)(x - 7)$

e $(x - 2)(x + 8)$

f $(x - 9)(x + 4)$

g $(x - 1)(x - 3)$

h $(x - 5)(x - 4)$

i $(x + 2)(x - 2)$

3 Expand

a $(2 + a)(3 - b)$

b $(p - 3q)(a + 2b)$

c $(2c - 5d)(3c - 4f)$

4 Expand and simplify

a $(y + 1)^2$

b $(3 - x)^2$

c $(2 - 5d)^2$

d $(3y + 2)^2$

e $(4a + x)^2$

f $(3c - 4d)^2$

5 Expand and simplify

a $(3a + b)(3b + a)$

b $(2x + 3y)(x - y)$

c $(4x - 3y)(2x - y)$

d $(3p - q)(3q - 2p)$

e $(x + y)(x - y)$

f $(c - d)(c + 2d)$

g $(2p - q)(2p + q)$

h $(2x + 5y)(2x - 3y)$

i $(5x - 4y)(5x + 4y)$

6 Expand and simplify $(2n - 3)^2 - 4(n - 1)^2$

Exercise 8D

1 Factorise completely

a $x(2y + 1) + 5(2y + 1)$

b $x(a - b) + 2y(a - b)$

c $2a(x + 3y) - 3b(x + 3y)$

d $y(x - 4) + 1(4 - x)$

e $(x + 7)^2 + 3(x + 7)$

f $(x - y)^2 - b(y - x)$

g $(x - 3y) + 2(3y - x)^2$

h $(2x + 3)(x + 10) + (x + 10)$

i $qx + xp + 3q + 3p$

j $xa - 2bx + ya - 2yb$

k $2x^2 - 4x + x - 2$

l $x^2 - 4x - 5x + 20$

m $4x^2 + 6x - 10x - 15$

n $6x^2 + 4 - 3x - 8x$

Exercise 8E

1 a Write down the two numbers whose product is $+24$ and whose sum is $+14$

b Hence factorise $x^2 + 14x + 24$

2 a Write down the two numbers whose product is -24 and whose sum is $+2$

b Hence factorise $x^2 + 2x - 24$

3 a Write down the two numbers whose product is $+24$ and whose sum is -11

b Hence factorise $x^2 - 11x + 24$

4 Factorise

a $x^2 + 13x + 12$ **b** $x^2 - 13x + 12$

c $x^2 + 8x + 12$ **d** $x^2 - 7x + 12$

e $x^2 + 11x - 12$ **f** $x^2 - 4x - 12$

g $x^2 + 11x + 18$ **h** $x^2 + 9x + 18$

i $x^2 - 7x - 18$ **j** $x^2 - 17x - 18$

k $x^2 + 12x + 36$ **l** $y^2 - 2y + 1$

m $x^2 - 9x - 10$ **n** $x^2 + 8x + 15$

o $x^2 + 16x - 17$ **p** $x^2 - 4x - 21$

q $x^2 + 20x + 64$ **r** $y^2 - 16y + 64$

Exercise 8F

1 Factorise

a $y^2 - 4$ **b** $4 - y^2$

c $25 - z^2$ **d** $p^2 - 121$

e $16 - t^2$ **f** $1 - z^2$

g $4x^2 - 49$ **h** $p^2 - 25x^2$

i $64 - 81q^2$ **j** $16x^2 - y^2$

k $(x + 2)^2 - 9$ **l** $49x^2 - (x + 5)^2$

m $100x^2 - (2 - 3x)^2$

2 a Factorise $m^2 - n^2$.

b Hence find the value of

  i $23^2 - 21^2$ **ii** $13.75^2 - 6.25^2$.

3 Factorise completely

a $2x^2 - 98$ **b** $75x^2 - 12$

c $45y^2 - 5x^2$ **d** $90a^2 - 40b^2$

e $5p^2 - 20q^2$ **f** $12(x + 1)^2 - 3x^2$

Chapter 9 Two-dimensional shapes

Exercise 9A

1 Use ruler and compasses to construct this triangle accurately. You must show all construction lines. Measure and write down the size of the smallest angle in the triangle.

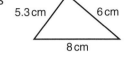

2 Use ruler and compasses to **construct** an equilateral triangle with sides of length 6 centimetres.

You must show all construction lines.

(1387 June 2005)

3 The diagram shows the triangle ABD.

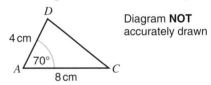

Diagram **NOT** accurately drawn

a Make an accurate drawing of the triangle ABD.

b C is the point so that $ABCD$ is a parallelogram.

Mark the position of C with a cross (**X**). Label the point C.

4 $ABCD$ is a rhombus of side 7 cm.

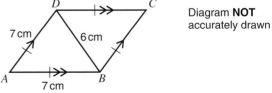

Diagram **NOT** accurately drawn

The length of the diagonal BD is 6 cm.

Use ruler and compasses to construct the rhombus $ABCD$.

You must show **all** construction lines.

(5540 June 2005)

Exercise 9B

1 Write down the name of two quadrilaterals with

  a four right angles

  b two pairs of adjacent sides equal in length

  c both pairs of opposite sides equal in length.

2 Here is a regular hexagon *ABCDEF*.

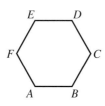

  Write down the mathematical name of each of these quadrilaterals.

  a *ABDE*   **b** *ABDF*   **c** *ABCF*.

3 The length of a rectangle is 12 cm and its width is 5 cm.
Work out

  a the perimeter   **b** the area.

4 The perimeter of a square is 28 cm.
Work out its area.

5 Work out the areas of these parallelograms.

  a

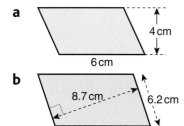

  b

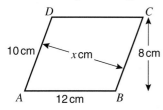

6 *ABCD* is a parallelogram.

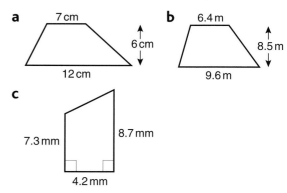

  AB = 12 cm and *AD* = 10 cm.
The distance between the parallel sides *AB* and *DC* is 8 cm.
The distance between the parallel sides *AD* and *BC* is *x* cm.
Work out the value of *x*.

7 The diagram shows a sketch of a triangle.

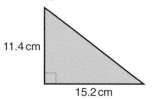

  Work out the area of the triangle.
State the units with your answer.

(1385 June 2000)

8 Work out the areas of these triangles.

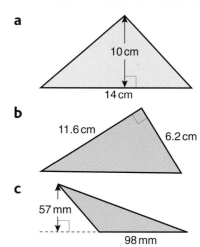

12 Work out the areas of these trapezia.

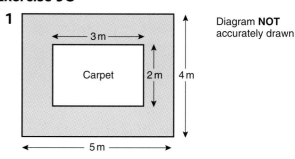

Exercise 9C

1

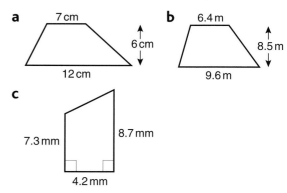

Diagram **NOT** accurately drawn

The diagram shows the plan of a floor.
There is a carpet in the middle of the floor.
Work out the shaded area.

(5540 June 2005)

2 The diagram shows a 5 m by 4 m rectangular pond.

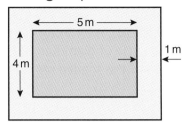

There is a path of width 1 m around the pond.
The inner edge of the path is a 5 m by 4 m rectangle.
Work out the area of the path.

3

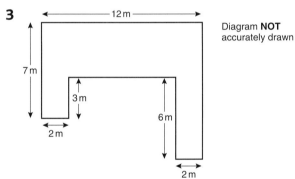

Diagram **NOT** accurately drawn

The diagram shows a paved surface.
All the corners are right angles.
Work out the area of the paved surface.

(1385 June 2002)

4 The diagram shows a 6-sided shape made from a rectangle and a right-angled triangle.

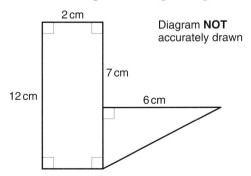

Diagram **NOT** accurately drawn

Work out the total area of the 6-sided shape.

(1387 November 2005)

5 A floor is a 5 m by 3.5 m rectangle.
Carpet tiles, which are squares of side 50 cm, are used to carpet the floor.
Work out how many carpet tiles are needed.

6 The length of each diagonal of a square is 20 cm.

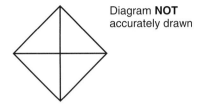

Diagram **NOT** accurately drawn

Work out the area of the square.

(1387 November 2004)

Exercise 9D

If your calculator does not have π button, take the value of π to be 3.142
Give answers correct to 3 significant figures unless stated otherwise.

1 Work out the circumferences of circles with these diameters.

    a 13 cm   **b** 31 mm   **c** 4.3 m
    d 8.4 cm   **e** 23.6 m

2 Work out the circumferences of circles with these radii.

    a 12 m   **b** 5.6 cm   **c** 11 mm
    d 6.74 m   **e** 5.6 mm

3 Work out the diameters of circles with these circumferences.

    a 37 mm   **b** 42 m   **c** 48.5 cm
    d 54.2 cm   **e** 67.36 m

4 A semicircle has a diameter of 7.8 cm.
Work out its perimeter.
(Hint: the perimeter includes the diameter.)

5 A quarter circle has a radius of 5.8 cm.
Work out its perimeter.
(Hint: the perimeter includes the two radii.)

6 A shape is made from 27 mm by 16 mm rectangle and a semicircle.

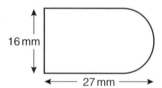

Work out its perimeter.

7 A shape is made from 8.7 cm by 4.8 cm rectangle and a quarter circle. Work out its perimeter.

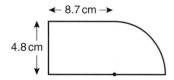

8 The diameter of a unicycle wheel is 60 cm. Work out the number of complete turns the wheel makes when the unicycle travels 500 metres.

Exercise 9E

If your calculator does not have π button, take the value of π to be 3.142
Give answers correct to 3 significant figures unless stated otherwise.

1 Work out the areas of circles with these radii.

   a 5 cm   **b** 4.7 m   **c** 3.9 mm

   d 1.9 cm   **e** 2.83 m

2 Work out the areas of circles with these diameters.

   a 9 cm   **b** 11 mm   **c** 7.3 m

   d 10.3 cm **e** 9.14 m

3 The radius of a semicircle is 3.4 cm. Work out its area.

4 The diameter of a semicircle is 7.5 cm. Work out its area.

5 The radius of a quarter circle is 9.3 m. Work out its area.

6 A shape is made from 4.3 m by 2.7 m rectangle and a semicircle.
Work out its area.

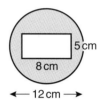

7 The diagram shows a 8 cm by 5 cm rectangle inside a circle of diameter 12 cm. Work out the area of the shaded region.

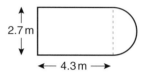

8 Swinside stone circle, in the Lake District, has a radius of 14.5 m.

   a Work out the circumference of the circle.

   b Work out the area enclosed by the circle correct to the nearest m².

Exercise 9F

In Questions **1–4**, give the answers as multiples of π.

1 Find the circumference of a circle with a diameter of 5 m.

2 Find the area of a circle with a radius of 7 cm.

3 Find the circumference of a circle with a radius of 10 cm.

4 Find the area of a circle with a diameter of 12 m.

5 The diameter of a semicircle is 6 cm. Find its perimeter. Give your answer in terms of π.

6 The radius of a semicircle is 8 cm. Find its area. Give your answer in terms of π.

7 The diameter of a quarter circle is 40 cm. Find its perimeter. Give your answer in terms of π.

8 The radius of a quarter circle is 6 cm. Find its area. Give your answer in terms of π.

9 The circumference of a circle is 24π cm. Find its diameter.

Exercise 9G

If your calculator does not have π button, take the value of π to be 3.142.
Give answers correct to 3 significant figures unless stated otherwise.

1 Calculate a the arc length and b the perimeter of each of these sectors.

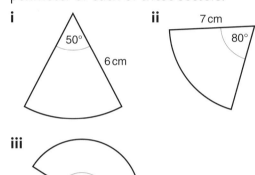

i

50°
6 cm

ii

7 cm
80°

iii

9 cm 100°

2 Calculate the area of each of the sectors in Question 1.

3 a Find the perimeter of each of these sectors. Give each answer in terms of π.

   b Find the area of each of these sectors. Give each answer as a multiple of π.

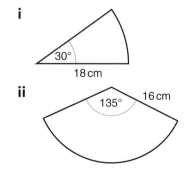

i

30°
18 cm

ii

135° 16 cm

4 Calculate the area of the shaded segment.

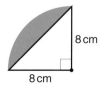

8 cm

8 cm

Exercise 9H

1 Change 8560 mm² to cm².

(1387 June 2004)

2 Change to cm²

   a 2 m²   **b** 8.4 m²   **c** 834 mm²

3 Change to mm²

   a 7 cm²   **b** 21 cm²   **c** 5.2 m²

4 Change to m²

   a 90 000 cm²   **b** 360 000 cm²

   c 5 km²

5 A rectangle measures 8.3 cm by 45 mm. Find the area of the rectangle in

   a cm²   **b** mm²

6 Work out the area of this triangle in

   a m²   **b** cm².

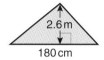

2.6 m

180 cm

Chapter 10 Linear equations

Exercise 10A

Solve these equations.

1 $4b = 52$	**2** $3e - 7 = 17$
3 $9 + 2m = 27$	**4** $21 = 2p - 9$
5 $4 = 8q + 4$	**6** $12 + 7y - 2 = 52$
7 $5c - 1 = 0$	**8** $6f - 24 = 9$
9 $8 + 3h = 15$	**10** $0 = 8m - 6$
11 $9 = 8 + 3p$	**12** $2 + 10y - 15 = 3$
13 $4b + 27 = 3$	**14** $2c + 9 = 4$
15 $9 - 4d = 33$	**16** $26 = 32 + 8k$
17 $0 = 15 + 2m$	**18** $3p - 4 = -13$
19 $2 - 5q = 13$	**20** $-3 - 7r = 16$
21 $21 = 8 - 10t$	**22** $2(x + 3) = 14$
23 $2(a - 1) = 9$	**24** $4(2b + 4) = 28$
25 $2(4d + 1) = 14$	**26** $2(3e + 7) = 17$
27 $4(2f - 3) = 2$	**28** $3(g + 2) = 10$
29 $2(1 - h) = 7$	**30** $5(3 - 4m) = 35$

31 $26 = 7 + 2(p+1)$

32 $3(4q + 1) - 7q = 10$

33 $6(x + 5) + x = 23$

34 $7x = 5x + 4$

35 $5y = 2y + 9$

36 $3z = z - 16$

37 $3a = 16 - a$

38 $2b = 8 - 3b$

39 $3c = 4 - 2c$

40 $3d + 1 = 2(d + 1)$

41 $2(e + 5) = e + 8$

42 $4(3f - 1) = 2f + 9$

43 $2(5g + 7) = 7g + 11$

44 $2(2h + 1) = 16 - 3h$

45 $5(k - 1) = 3k - 19$

46 $3(2m - 1) = 4(m - 1)$

47 $3(n + 2) = 2(n + 3)$

48 $2(4p - 7) = 2(1 - p)$

Exercise 10B

1 Brendan thinks of a number. He multiplies the number by 6 then adds 17. His answer is 59. Work out the number that Brendan thinks of.

2 Four buses took 200 people from Middleton to Blackpool last Tuesday.
The first three buses each had x people.
The fourth bus had 20 less people than each of the other buses.

   a Write down an expression, in terms of x, for the number of people on the fourth bus.

   b By forming an equation, work out the value of x.

3 The width of a rectangle is x centimetres. The length of the rectangle is $(x + 4)$ centimetres.

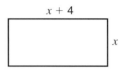

   a Find an expression, in terms of x, for the perimeter of the rectangle.
Give your expression in its simplest form.

The perimeter of the rectangle is 54 centimetres.

   b Work out the length of the rectangle.
(1388 June 2005)

4

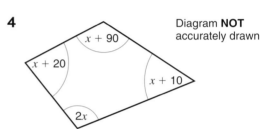

Diagram **NOT** accurately drawn

The sizes of the angles, in degrees, of the quadrilateral are

$x + 10$ $2x$ $x + 90$ $x + 20$

   a Use this information to write down an equation in terms of x.

   b Use your answer to part **a** to work out the size of the smallest angle of the quadrilateral.
(1387 November 2005)

5 The lengths, in centimetres, of the sides of a quadrilateral are

$2x$ $x + 6$ $4x - 12$ $30 - 3x$

   a Explain why x must be
     i greater than 3   **ii** less than 10.
The perimeter of the quadrilateral is 48 cm.

   b Use this information to write down an equation in terms of x.

   c Find the value of x.

   d The angles of the quadrilateral are not all equal. Explain why the quadrilateral is a rhombus.

6

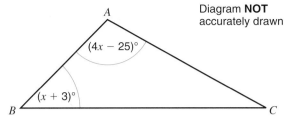

Diagram **NOT** accurately drawn

ABC is a triangle
Angle $A = (4x - 25)°$.
Angle $B = (x + 3)°$.
The size of angle A is **three** times the size of angle B.
Work out the value of x.

(1388 November 2005)

7 Hassan, Joanne and David are selling their cars. Hassan's car has done 2550 miles more than Joanne's car and David's car has done 1500 miles less than Joanne's car. Joanne's car has done x miles.

a Write down an expression, in terms of x, for the number of miles David's car has done.

b Hassan's car has done four times as many miles as David's car.
By forming an equation, find the value of x.

c How many miles has Hassan's car done?

Exercise 10C

1 Solve these equations.

a $\dfrac{a}{2} = 3$ **b** $\dfrac{24}{b} = 8$ **c** $\dfrac{c}{4} = 0.5$

d $\dfrac{6}{d} = 12$ **e** $\dfrac{24}{e} = 32$ **f** $\dfrac{36}{f} = 24$

2 Solve these equations.

a $\dfrac{x + 2}{2} = 5$ **b** $\dfrac{y - 6}{4} = 7$

c $\dfrac{x}{5} - 2 = 1$ **d** $2 = \dfrac{p - 7}{7}$

e $7 = \dfrac{2q - 3}{3}$ **f** $\dfrac{2p + 1}{2} = 1$

g $\dfrac{3x - 2}{2} = 3$ **h** $\dfrac{2x}{5} - 2 = 3$

i $\dfrac{2 + 5y}{3} = 3$ **j** $7 + \dfrac{x}{2} = 3$

k $\dfrac{3 - y}{3} = 3$ **l** $\dfrac{2z + 21}{5} = 2$

m $x - 2 = \dfrac{x}{2}$ **n** $\dfrac{2y}{3} = y - 5$

o $8 - \dfrac{3x}{5} = x$ **p** $\dfrac{x}{2} + \dfrac{x}{4} = 15$

q $\dfrac{9x + 1}{12} + \dfrac{x - 4}{3} = 2$

r $\dfrac{2y - 3}{3} + \dfrac{4 - 3y}{6} = \dfrac{7y - 5}{6}$

s $\dfrac{x + 2}{8} + \dfrac{2 - 3x}{4} = \dfrac{9}{2}$

t $\dfrac{4x + 1}{6} + \dfrac{2x + 1}{4} = \dfrac{4x + 5}{2}$

3 Solve $2(x + 5) = \dfrac{8x - 5}{3}$

(1388 March 2006)

4 Karen thinks of a number, n. She multiplies the reciprocal of the number by 4 then subtracts her answer from 1 to get $\dfrac{3}{5}$.

a Write down an equation in terms of n.

b Solve the equation to find n.

5 Jim drives for 30 minutes at an average speed of x miles per hour. He then drives for 45 minutes at an average speed of $(2x - 26)$ miles per hour. He drives a total distance of 51 miles. Work out the value of x.

Exercise 10D

1 Solve the simultaneous equations
$5a + 3b = 9$
$2a - 3b = 12$

(1387 June 2006)

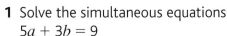

2 Solve the simultaneous equations
$3x + y = 1$
$x - 2y = 19$

(1388 November 2005)

3 Solve these simultaneous equations.

a $x + 2y = 9$
$x + y = 7$

b $3x + y = 23$
$y = x + 3$

c $3p + 2q = 8$
$5p - 2q = 16$

d $x + y = 9$
$y = 1 - 2x$

e $6x + 4y = 35$
$2x - 3y = 3$

f $x + 2y = 37$
$y = 3x + 1$

g $3x + 4y = 5$
$2y = 3x - 2$

h $y - 6x = 12$
$3y - 10x = 16$

4 Solve the simultaneous equations
$3x + 7y = 26$
$4x + 5y = 13$

(1387 November 2005)

Exercise 10E

1 8 cakes and 3 pies cost a total of £6.40
5 cakes and 2 pies cost a total of £4.05

a Work out the cost of a cake.

b Work out the total cost of a cake and a pie.

2 The diagram shows an equilateral triangle.

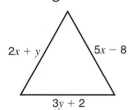

All sides are measured in centimetres.

a Show that $3y = 5x - 10$

b Find the value of x and the value of y.

c Find the perimeter of the triangle.

3 The sum of two numbers is 24.
The larger number is two more than three times the smaller number.
Find the larger number.

4 A hotel has single and double rooms.
A single room is for one person and a double room is only for two people.
The cost of staying in a single room is £80 per night.
The cost of staying in a double room is £60 per person per night.
51 people stayed at the hotel last night for a total cost of £3800.
Work out the number of people who stayed in the single rooms.

5 Tickets for a cricket match cost £36 for an adult and £20 for a child.
20 000 tickets are sold for a total of £665 600

a Work out the number of adult tickets sold.

b Find the total amount paid for the child tickets.

Chapter 11 Collecting and recording data

Exercise 11A

1 Write down whether each of the following data is qualitative data or quantitative data.

a The heights of garden walls.

b The colours of cars.

c The weights of potatoes.

d The volumes of beakers.

e The sizes of shoes.

f The days of the week.

2 Write down whether each of the following data is discrete data or continuous data.

a The scores on a dartboard.

b The numbers of CDs owned

c The numbers of GCSE passes.

d The times taken to travel to school.

e The heights jumped at the high jump.

f The sizes of dresses.

3 Jamilla spins a spinner 50 times.
Here are the scores on the spinner for each of the 50 spins.

```
1  3  2  3  4  2  3  3  2  3
2  1  3  4  2  3  2  2  1  2
3  4  2  3  3  3  2  1  4  2
2  2  2  1  2  3  1  3  4  3
4  1  2  3  3  2  2  2  3  4
```

a Draw a tally chart to show this information.

b Is the spinner fair or biased? Explain your answer.

4 20 men and 18 women were asked their age.
6 men were over 35 years of age.
15 people, of which 6 were women, were from 26 to 35 years of age.
2 women were from 18 to 25 years of age.
Copy and complete the two-way table.

	18 to 25 years	26 to 35 years	Over 35	Total
Men				
Women				
Total				38

5 Zach recorded the number of chapters in some books.
Here are his results.

```
24  43  18  26  36  42  28  34  32  40
19  21  31  43  25  42  27  36  29  30
21  31  46  24  28  41  37  29  30  27
```

a Copy and complete the grouped frequency table.

Number of chapters	Tally	Frequency
11–20		
21–30		
31–40		
41–50		

b Write down the modal class interval.

6 The table shows some information about the weights, in grams, of some apples

Weight (w grams)	Frequency
$180 \leqslant w < 200$	15
$200 \leqslant w < 220$	24
$220 \leqslant w < 240$	18
$240 \leqslant w < 260$	32
$260 \leqslant w < 280$	17

a Write down the modal class.

b Work out the total number of apples.

c Find the number of apples whose weight was less than 240 grams.

d Find an estimate for the number of apples whose weight was less than 210 grams.

Exercise 11B

1 A student wanted to find out how many pizzas adults ate. He used this question in a questionnaire.

> How many pizzas have you eaten?
>
> ☐   ☐
> A few   A lot

a Write down **two** things that are wrong with this question.

b Design a better question that the student can use to find out how many pizzas adults ate.
You should include some response boxes.

2 Write down two things wrong with each of the following questions.

a "How much television do you watch?"

☐   ☐   ☐
not much   average   lots

b "How much pocket money do you get each week?"

£1–£2   £2–£3   £3–£4   £4–£5
☐   ☐   ☐   ☐

3 Michelle wants to find out how often people visit the local library .

  a Design a suitable question to find out how often people visit the local library.

Michelle decides to use a sample of 100 people.
She stands by the entrance to the library on a Monday morning and gives a questionnaire to each of the first 100 people that she meets.

  b **i** What is wrong with this sample?
     ii How could Michelle get a better sample?

4 The two-way table shows the age range and gender of 200 members of an angling club. A stratified sample of 40 members, stratified by age and gender, is to be surveyed on the condition of the local rivers.

	11–17	18–30	31–55	Over 55	Total
Male	13	40	58	29	140
Female	5	16	25	14	60
Total	18	56	83	43	200

  a Work out the number of members who are male and in the age range 18–30 in the stratified sample.

  b Work out the number of members who are female and in the age range 31–55 in the stratified sample.

  c Work out the number of members in the age range 11–17 in the stratified sample.

5 A newspaper employs 2400 people. Describe two ways in which you could choose a random sample of 10% of these people.

Exercise 11C

1 The following database contains some information about the people living in a street Using this database,

House number	Number of occupants	Number of pets	Number of cars
1	3	2	2
2	6	0	3
3	2	4	1
4	3	2	3
5	4	1	2

  a write down the number of pets at house number 3.

  b write down the number of occupants at house number 2.

  c write down the total number of houses with more than 1 car.

  d write down the total number of pets in the five houses.

2 This database contains information about some students.

Name	Gender	Age	IQ
Jessica	female	12	101
Daniel	male	11	112
Mason	male	13	102
Samantha	female	16	98
Melissa	female	13	105
Zach	male	13	97
Georgina	female	15	95
Marc	male	16	100
Alfie	male	12	104
Helen	female	13	120

Using the database,

  a write down Zach's age.

  b write down Melissa's IQ.

  c write down the name of the oldest female.

  d who has an IQ under 100?

  e who has the highest IQ?

  f how many students are under 14 years of age?

  g list the girls in order of IQ highest IQ first.

3 The database opposite contains some information about the cost (in £) of holidays at a Spanish hotel.
Using this database,

  a write down the cost of a holiday for 14 nights in June.

  b write down the cost of a holiday for 4 nights in May.

  c write down the holidays which cost over £1000

  d write down a list of all of the holidays

Month	4 nights	7 nights	10 nights	14 nights
April	385	495	619	775
May	499	615	755	919
June	465	605	755	935
July	535	705	895	1105
August	605	775	945	1169
September	535	675	829	999
October	525	685	799	949

4 Joanna wants to find out if females with blue eyes watch more television than other females.

Name	Age	Month of birth	Gender	Colour of eyes	Favourite colour	Favourite sport	Favourite Capital City	TV hours watched per week	Number of pets
Angela	29	April	Female	Blue	Red	Football	Amsterdam	25	1
Bryony	16	June	Female	Green	Black	Tennis	London	34	4
Caroline	35	December	Female	Blue	Green	Pool	Madrid	18	7
Doris	13	May	Female	Brown	Blue	Baseball	Paris	22	3
Elaine	21	September	Female	Green	White	Rugby	London	42	0

Joanna does not need to use all of this database.
Which parts of the database should she use?

Chapter 12 Percentages

Exercise 12A

1 Write each percentage as a fraction.

  a 67% **b** 3% **c** 39% **d** 13%

2 Write each percentage as a fraction in its simplest form.

  a 6% **b** 60% **c** 35% **d** 24%

3 Write each percentage as a decimal

  a 46% **b** 71% **c** 5%

  d 30% **e** 14.5% **f** 3.2%

4 18% of children go by train to school.
What fraction of children go by train to school.
Give your fraction in its simplest form.

5 95% of households own a television.
Write down the fraction of households that does **not** own a television.
Give your fraction in its simplest form

6 Write each percentage as: **i** a fraction in its simplest form, **ii** a decimal.

  a $37\frac{1}{2}$% **b** $5\frac{1}{2}$% **c** $7\frac{1}{4}$%

7 Work out

  a 50% of 40   **b** 25% of 20

  c 10% of 80   **d** 50% of 24

  e 20% of 50   **f** 25% of 48

  g 75% of 16   **h** 75% of 80

8 Work out

  a 10% of £400   **b** 50% of £280

  c 25% of 40 g   **d** 75% of 200 m

  e 2% of £300   **f** 45% of 400kg

  g 12% of $200   **h** 5% of £420

9 Mikhail invests £400. The interest rate is 4% per year. How much interest will he receive after one year?

10 There are 600 students at Northolt School. 55% of these 600 students are girls. How many of the students are girls?

11 Work out
 a 24% of £30   b 5% of 45.6 m
 c 80% of 65 kg   d 6.5% of £300
 e 4.2% of £2500 f 17½% of £560
 g 2¼% of 30 m

12 Alex invests £6200. The interest rate is 4.1% per year. How much interest will he receive at the end of 1 year?

13 Bhavana did a maths test. There was a total of 60 marks for the test. Bhavana got 45% of the marks. Work out how many marks she got.

14 Ezra earns £325 per week. He spends 42% of his weekly wage on rent. Work out how much Ezra spends on rent.

15 There are 24 600 books in a library. 46% of the books in the library are fiction. Work out how many fiction books there are in the library.

16 62% of the workers in a factory are female. There are 850 workers at the factory. Work out how many of the workers are female.

Exercise 12B

1 Write down the multiplier that can be used to work out an increase of
 a 37% b 8%   c 13%   d 30%
 e 2.7% f 17½% g 200% h 175%

2 Write down the multiplier that can be used to work out a decrease of
 a 4%   b 15% c 30%   d 21%
 e 4.3% f 17½% g 0.5% h 1¼%

3 Janet earns £400 per week. She gets a wage rise of 10%. How much does Janet earn per week after her rise?

4 In a sale, prices are reduced by 15%. Work out the sale price of a washing machine that normally costs £360.

5 Alan puts £800 into a bank account. At the end of one year 4% interest is added. How much is in her account at the end of one year?

6 A holiday normally costs £780. It is reduced by 17.5%. How much will the holiday now cost?

7 William's salary is £24 000
His salary increases by 4%.
Work out William's new salary.
(1388 March 2005)

8 The price of a DVD player was £120
In a sale, the price is reduced by 35%.
Work out the sale price of the DVD player.
(1388 November 2005)

9 Alistair sells books.
He sells each book for £7.60 plus VAT at 17½%.
He sells 1650 books
Work out how much money Alistair receives.
(1387 June 2005)

Exercise 12C

1 Write as a percentage.
 a £5 out of £20
 b 3 m out of 6 m
 c 6 cm out of 60 cm
 d 80 kg out of 100 kg
 e $40 out of $200
 f 21 g out of 28 g

2 Write as a percentage.
 a 50p out of £2
 b 80 cm out of 4 m
 c 300 m out of 1 km
 d 24 minutes out of 1 hour
 e 45 mm out of 6 cm
 f 80p out of £4

3 Linda's mark in a maths test was 36 out of 50. Find 36 out of 50 as a percentage.
(1388 January 2005)

4 Lesley's mark is a geography test was 48 out of 60.
Work out 48 out of 60 as a percentage.
(1388 March 2002)

5 Frank buys a crate of 60 apples. He finds that 9 of the apples are rotten.

   a What percentage of the apples are rotten?

   b What percentage of the apples are **not** rotten?

6 Jalin was given £600. He put £420 out of the £600 into a savings account. What percentage of the £600 did Jalin put into a savings account?

7 Digicam is a shop that sells digital cameras. In 2004, Digicam sold 920 digital cameras. In 2005, Digicam sold 1050 digital cameras. Work out the percentage increase in the number of digital cameras sold. Give your answer correct to 2 significant figures.

8 Jack buys a box of 20 pens for £3.00 He sells all the pens for 21p each. He sells all the pens. Work out his percentage profit.
(1385 November 2002)

9 Sue buys a pack of 12 cans of cola for £4.80. She sells **all** the cans for 50p each. She sells all of the cans. Work out her percentage profit.
(1385 June 2000)

10 Yasmin opened an account with £750 at a bank. After one year, the bank paid her interest. She then had £776.25 in her account. Work out, as a percentage, the bank's interest rate.

11 In 2000 the value of a house was £78 000. In 2006 its value was £105 300.

   a Work out the value of the house in 2006 as a percentage of the value of the house in 2000.

   b If the index number in 2000 is 100, write down the index number in 2006.

12 The following table shows the cost of the same car in some of the years from 2002 to 2005. Some of the index numbers, based on an index of 100 in 2002, are also shown.

Year	2002	2003	2004	2005
Cost of car	12 000	12 600	12 900	
Index number	100		107.5	114

   a Work out the index number in 2003.

   b Work out the price of the car in 2005.

Exercise 12D

1 Work out the multiplier as a single decimal number that represents

   a an increase of 15% for 2 years

   b a decrease of 12% for 3 years

   c an increase of 4.5% for 4 years

   d an increase of 10% for 5 years

   e a decrease of 23% for 2 years

   f an increase of 25% followed by a decrease of 6%

   g a decrease of 11% followed by an increase of 15%

   h an increase of 25% followed by a decrease of 25%

2 £2400 is invested for 2 years at 4% per annum **compound interest**. Work out the total amount in the account after 2 years.

3 £6000 is invested for 3 years at 5% per annum **compound interest**. Work out the **total interest** earned over the 3 years.

4 A car is worth £14 500. Each year the value of the car depreciates by 28%. Work out the value of the car after 3 years.

5 Each year the value of a cooker falls by 9% of its value at the beginning of that year. Imogen bought a new cooker for £620. Work out the value of the cooker after 3 years.

6 Bill buys a new machine. The value of the machine depreciates by 20% each year.

   a Bill says 'after 5 years the machine will have no value'.

   Bill is wrong. Explain why.

Bill wants to work out the value of the machine after 2 years.

b By what single decimal number should Bill multiply the value of the machine when new?

(1387 November 2005)

7 Henry invests £4500 at a compound interest rate of 5% per annum.
At the end of n complete years the investment has grown to £5469.78
Find the value of n.

(1387 November 2003)

Exercise 12E

1 Max receives a pay increase of 5%.
Max now earns £231 per week.
Work out Max's weekly wage before the increase.

2 In a sale, normal prices are reduced by 15%.
The sale price of a CD player is £102
Work out the normal price of the CD player.

(1388 March 2006)

3 Jacob answered 80% of the questions in a test correctly.
He answered 32 of the questions correctly.
Work out the total number of questions in the test.

(1388 January 2004)

4 A clothes shop has a sale.
All the original prices are reduced by 24% to give their sale price.
The sale price of a jacket is £36.86
Work out the original price of the jacket.

(1385 November 1999)

5 Amos opened a bank account. The bank's interest rate was 4.5%. After one year, the bank paid him interest. The total amount in the account was then £1254. Work out the amount with which he opened his account.

6 Sunita invests some money in a bank account.
Compound interest is paid at a rate of 4.2% per annum. After 3 years there is £3167.83 in the account. How much did Sunita invest?

Chapter 13 Graphs (1)

Exercise 13A

1 Here is part of a travel graph of Siân's journey from her house to the shops and back.

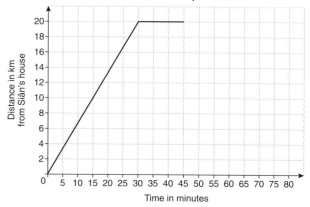

a Work out Siân's speed for the first 30 minutes of her journey.
Give your answer in km/h.

Siân spends 15 minutes at the shops. She then travels back to her house at 60 km/h.

b Complete the travel graph.

(1387 June 2003)

2 The graph shows Becky's journey to see her friend who lives 18 km from Becky's home.

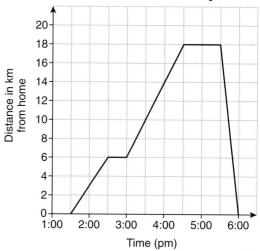

a At what time did Becky leave home?

b At what time did she arrive at her friends?

c Becky stopped on the way to see her friend.
  i How is this shown on the graph?
  ii For how long did she stop on the way?

d How many minutes did the return journey take?

e Work out the average speed, in kilometres per hour, of each part of Becky's journey.

3 The diagram represents the velocity–time graph of part of a train journey.

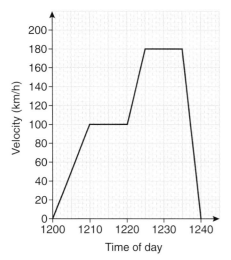

a Write down the maximum velocity of the train.

b Find the constant acceleration of the train during the first 10 minutes of the journey.

c Find the constant acceleration of the train from 1220 to 1225.

d For how many minutes did the train have zero acceleration?

e Describe fully the journey of the train from 1225 to 1240.

4 The diagram shows a container filling up with water.

d is the diameter of the surface of the water when the height of the water is h. Sketch a graph to show the relationship between d and h.

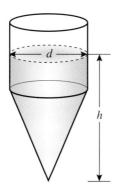

Exercise 13B

1 a Find the equation of the line which passes through the points $A(4, 6)$, $B(1, 3)$, $C(0, 2)$, $D(-2, 0)$ and $E(-4, -2)$

b Find the equation of the line which passes through the points $V(4, 20)$, $W(3, 15)$, $X(0, 0)$ and $Y(-2, -10)$

2 a Draw the graph of $y = 2x + 2$ taking values of x from -2 to 3.

b Write down the coordinates of the point where your graph crosses **i** the y-axis, **ii** the x-axis.

c Use your graph to find the value of x when $y = 5$

3 a Draw the graph of $y = 5x - 3$, for values of x from -1 to 3.

b i On the same axes draw the graph of $x = 1$

ii Write down the coordinates of the point where the two graphs cross.

c i On the same axes draw the graph of $y = 5$

ii Write down the coordinates of the point where the graph of $y = 5x - 3$ crosses the graph of $y = 5$.

4 a Draw the graph of $y = \frac{1}{2}x + 1$. Use values of x from $x = -4$ to $x = 4$.

b Find the coordinates of the point of intersection of $y = \frac{1}{2}x + 1$ and the line $x = -2$.

c The point with coordinates $(2, k)$ lies on the graph of $y = \frac{1}{2}x + 1$. Use your graph to find the value of k.

5 a On the same axes, draw the graphs of
i $x + y = 1$　　**ii** $x + y = 4$
iii $x + y = 0$　　**iv** $x + y = -2$

b What do you notice about the four graphs you have drawn?

Exercise 13C

For each question, draw two linear graphs on the same grid (each axis scaled from -5 to 5) to solve the simultaneous equations

1 a $y = 3x - 1$
　　$x + y = 3$

b $y = 3x - 1$
　$y = 0.5x - 1$

2 a $x + y = 4$
　　$y = x$

b $x + y = 4$
　$y = 2x + 1$

3 $x + y = 2$
　$y = 4x - 3$

4 $y = x + 2$
　$y = -0.5x - 1$

5 $y = 1 - 3x$
　$y = x - 5$

Exercise 13D

In **Questions 1–5** for each of the lines write down **a** the gradient **b** the y-intercept

1 $y = 5x + 8$

2 $y = 2x - 3$

3 $y = -4x$

4 $y = \frac{3}{4}x - 7$

5 $y = 2 - 5x$

In **Questions 6–10** find **a** the gradient and **b** the y-intercept of the line with equation

6 $3y = 5x + 1$

7 $x + y = 9$

8 $y = 3(3 - 4x)$

9 $2y + 5x = 20$

10 $3x - 2y + 6 = 0$

11 A line passes through the point $(0, -2)$ and has gradient 5. Find the equation of the line.

12 a Find the gradient of the line joining the points $(0, 2)$ and $(2, 8)$.

b Find the equation of the line joining the points $(0, 2)$ and $(2, 8)$.

Exercise 13E

In **Questions 1–3** find the equation of the line which passes through the point A and has the given gradient

1 $A(0, 5)$ gradient 2

2 $A(1, -3)$ gradient -1

3 $A(4, -5)$ gradient 0.5

In **Questions 4–6** find an equation of the line which passes through the point P and is parallel to the line **L**

4 $P(0, -2)$ **L** $y = 2x$

5 $P(4, 0)$ **L** $y = 1 - 2x$

6 $P(-2, -5)$ **L** $2y = 3x + 4$

In **Questions 7–9** find the equation of the line which passes through the point Q and is perpendicular to the line **M**

7 $Q(2, 0)$ **M** $y = -\frac{1}{4}x + 7$

8 $Q(-1, 3)$ **M** $y = 2x + 3$

9 $Q(-1, -4)$ **M** $2y = 6x - 5$

In **Questions 10–12** find the equation of the line which joins the given points.

10 $(2, 3)$ and $(3, 7)$

11 $(-1, 0)$ and $(6, -1)$

12 $(-3, -1)$ and $(-1, 4)$

13

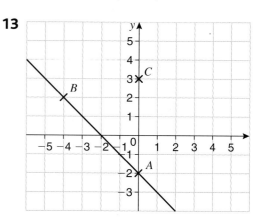

In the diagram A is the point $(0, -2)$, B is the point $(-4, 2)$, C is the point $(0, 3)$. Find an equation of the line that passes through C and is parallel to AB.

(1387 June 2006)

Chapter 14 Transformations

Exercise 14A

1 a Make a copy of shape **T** in the middle of a grid of centimetre squares.

b On your grid, translate shape **T**
  i 3 to the right and 2 up and label your new shape **A**,
  ii 2 to the right and 4 down and label your new shape **B**,
  iii by the vector $\begin{pmatrix} 1 \\ 4 \end{pmatrix}$ and label your new shape **C**,
  iv by the vector $\begin{pmatrix} -2 \\ 6 \end{pmatrix}$ and label your new shape **D**,
  v by the vector $\begin{pmatrix} -4 \\ -3 \end{pmatrix}$ and label your new shape **E**.

2

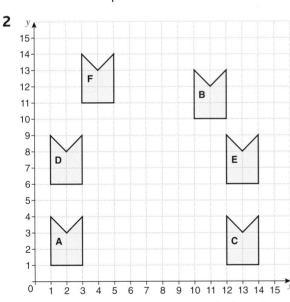

a Write down the vector which describes the translation that maps
  i shape **A** onto shape **B**, **ii** shape **A** onto shape **C**, **iii** shape **A** onto shape **D**,
  iv shape **F** onto shape **E**, **v** shape **C** onto shape **D**, **vi** shape **F** onto shape **D**.

b What is true about the translation that maps shape **D** onto shape **A** and the translation that maps shape **E** onto shape **C**?

c What is true about the translation that maps shape **D** onto shape **A** and the translation that maps shape **C** onto shape **E**?

Exercise 14B

1 Copy the diagram and rotate the triangle a half turn about the point O.

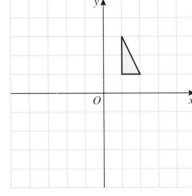

(1388 April 2005)

2

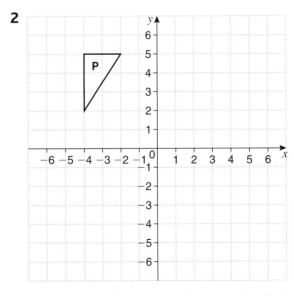

a Copy the diagram and rotate triangle **P** 90° clockwise about the point $(0, 2)$. Label the new triangle **Q**.

b On the same diagram, translate triangle **P** by the vector $\begin{pmatrix} 5 \\ -6 \end{pmatrix}$. Label the new triangle **R**.

(1388 November 2005)

3 Copy the diagram and rotate triangle **P** 90° anti-clockwise about the point (4, 2)

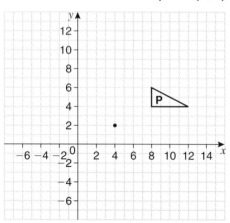

(4400 November 2005)

4 Describe fully the **single** transformation that maps **P** onto **Q**.

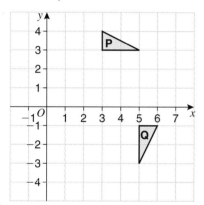

(4400 November 2005)

Exercise 14C

1 Copy the diagram and reflect the triangle in the *x*-axis.

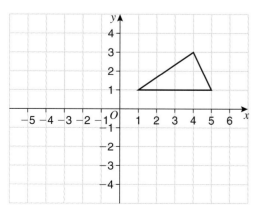

(1388 April 2005)

2 Triangle **A** and triangle **B** have been drawn on the grid.

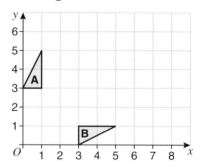

a Copy the diagram and
  i reflect triangle **A** in the line $x = 3$. Label this image **C**.
  ii reflect triangle **B** in the line $y = 2$. Label this image **D**.

b Describe fully the single transformation which will map triangle **A** onto triangle **B**. *(1387 June 2005)*

3 Triangle **B** is a reflection of triangle **A**.

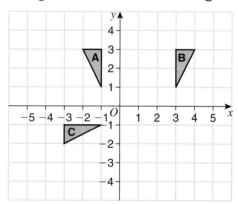

a **i** Copy the diagram and draw the mirror line for this reflection.
  ii Write down the equation of the mirror line.

b Describe fully the single transformation that maps triangle **A** onto triangle **C**.
(1385 November 2002)

Exercise 14D

1

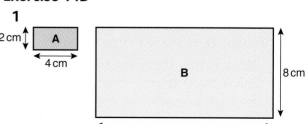

Rectangle **B** is an enlargement of rectangle **A**.

a Work out the scale factor of the enlargement.

b Work out the value of x

2 Copy the diagram and draw an enlargement, scale factor 3, of the shaded shape.

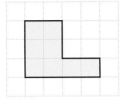

(1388 January 2004)

3 Copy the diagram and draw an enlargement, scale factor $\frac{1}{2}$, of the shaded shape.

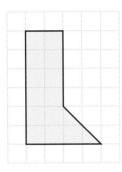

Exercise 14E

1 Copy of this diagram and enlarge the shaded triangle by a scale factor 2, centre O.

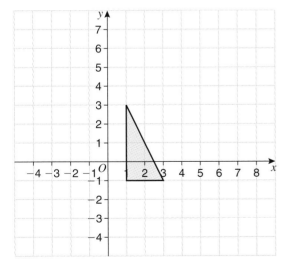

(1387 June 2004)

2 i Describe fully the single transformation that maps shape **P** onto shape **Q**.

ii Copy the diagram and reflect shape **P** in the line $x = 1$.

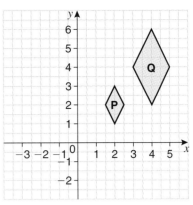

(1387 November 2004)

3 Copy the diagram and enlarge triangle **P** with scale factor $\frac{1}{2}$ and centre (4, 2).

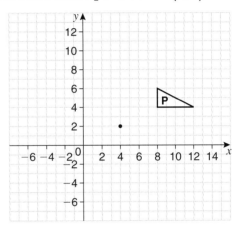

(4400 November 2005)

4

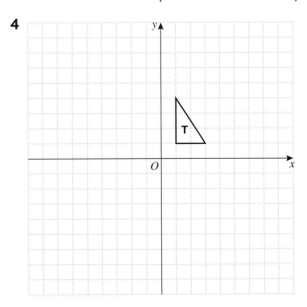

Copy the diagram and enlarge triangle **T**, scale factor -2, centre O.

(1388 January 2005)

5

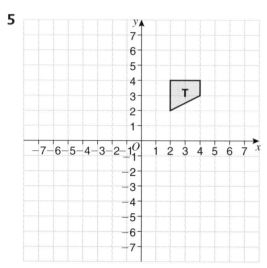

Copy the diagram and enlarge shape **T** with scale factor -1.5, centre $(0, 2)$.

(1388 March 2003)

Triangle **P** has been rotated $180°$ about the point $(1, 1)$ to give triangle **Q**.

a Copy the diagram and rotate triangle **Q** $180°$ about the point $(3, -1)$.
Label the triangle **R**.

b Describe the **single** transformation that takes triangle **P** to triangle **R**.

(1388 March 2004)

2

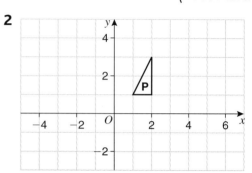

a Copy the diagram and reflect triangle **P** in the line $x = 3$.
Label the new triangle **Q**.

b Translate triangle **Q** by the vector $\begin{pmatrix} -8 \\ 0 \end{pmatrix}$.
Label the new triangle **R**.

c Describe fully the single transformation which maps triangle **P** onto triangle **R**.

6

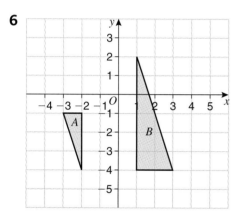

The diagram shows an enlargement from A to B.

i Write down the scale factor of the enlargement.

ii Find the coordinates of the centre of enlargement. *(1388 January 2002)*

Exercise 14F

1

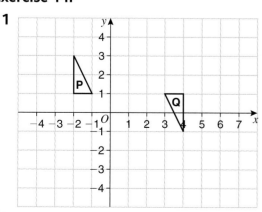

3

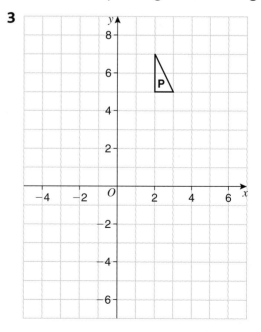

a Copy the diagram and reflect triangle **P** in the line in the line $y = 1$.
Label the new triangle **Q**.

b Reflect triangle **Q** in the line $y = -x$. Label the new triangle **R**.

c Describe fully the single transformation which maps triangle **P** onto triangle **R**.

Chapter 15 Inequalities

Exercise 15A

1 Write down the inequality shown on the number line:

a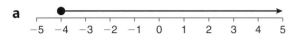

b

c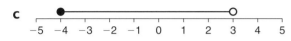

d

e

2 Show these inequalities on a number line.

a $x < -2$ **b** $x \geqslant -1$

c $-3 < x < -1$ **d** $0 \leqslant x \leqslant 2$

e $-1 \leqslant x < 5$

Exercise 15B

1 Solve these inequalities.

a $x - 3 < 6$ **b** $x + 1 > 4$

c $3x \leqslant 18$ **d** $6x \geqslant 9$

e $2x + 13 > 9$ **f** $3x - 5 < 7$

g $8x + 15 \leqslant 3$ **h** $10x - 5 \geqslant 0$

i $8x > 2x + 14$ **j** $7x \geqslant 20 + 3x$

k $3x \leqslant 4 - 5x$ **l** $5x - 3 \geqslant 9 - 5x$

2 a Solve the inequality $8x + 7 \leqslant 16 - 10x$.

b Show your solution on a number line.

3 Solve the inequality $6x < 7 + 4x$.
(1388 March 2006)

4 Solve these inequalities.

a $-3x < 12$ **b** $\dfrac{x - 3}{2} > 4$

c $3(x - 2) \leqslant 18$ **d** $6x \geqslant 2(4 + x)$

e $2(x + 3) \geqslant 5x - 9$ **f** $\dfrac{3x - 7}{4} < 2$

g $15 - 8x \leqslant 3$ **h** $x > \dfrac{3x}{5} + 4$

i $7 - 3(x - 2) \geqslant 2x + 17$

j $\dfrac{3 - x}{2} < 4 - \dfrac{x}{3}$

k $3(2x - 3) \leqslant 4 - 2(1 - 5x)$

Exercise 15C

1 $3 \leqslant x < 6$
x is an integer.
Write down all the possible values of x.

2 $-5 \leqslant y \leqslant -3$
y is an integer.
Write down all the possible values of y.

3 $-3 \leqslant z < 2$
z is an integer.
Write down all the possible values of z.

4 $-4 < 2x \leqslant 6$
x is an integer.
Write down all the possible values of x

5 $-3 < y - 2 < 2$
y is an integer.
Write down all the possible values of y.

6 $6a - 3 \leqslant 21$
a is a **positive** integer.
Write down all the possible values of a.

7 $3b + 10 > 1$
b is a **negative** integer.
Write down all the possible values of b.

8 a List all the possible integer values of n such that $-2 \leqslant n < 3$.

b Solve the inequality $4p - 8 < 7 - p$
(1387 June 2006)

9 a Solve $-3 \leqslant 2x - 7 \leqslant 10$

b x is an integer such that $-3 \leqslant 2x - 7 \leqslant 10$. Find all the possible values of x.

10

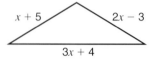

The diagram shows a triangle.
The lengths of the sides are given in centimetres.
The perimeter of the triangle is no more than 51 cm.

a Show that $1.5 < x \leqslant 7.5$

b Can the triangle be isosceles? Explain your answer.

Exercise 15D

1 The diagram shows the lines with equations $y = -2x + 4$ and $y = 4x - 5$

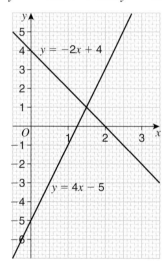

a From the graph, write down the solution of the simultaneous equations
$y = -2x + 4$
$y = 4x - 5$

b On a copy of the diagram, shade the region of points whose coordinates satisfy the four inequalities $y \leqslant -2x + 4$, $y \geqslant 4x - 5$, $2y - 6 < 0$ and $x \geqslant 0$.

c x and y are integers. Write down the coordinates of the set of points which satisfy all the four inequalities.

2 a Use a grid scaled from -4 to $+4$ on each axis to shade the region of points whose coordinates satisfy all the inequalities
$-4 < x \leqslant 3$, $y < 1$, $y \geqslant x$ and $3y + x \leqslant 3$.

b x and y are integers. Write down the coordinates of the set of points which satisfy all the inequalities $-4 < x \leqslant 3$, $y < 1$, $y \geqslant x$ and $3y + x \leqslant 3$.

3 a Show that the points $(0, 2)$ and $(4, 0)$ lie on the line with equation $x + 2y = 4$

b

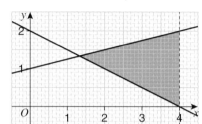

The diagram shows a shaded region bounded by three lines. Write down the three inequalities which must be satisfied by the coordinates of all points in the shaded region.

c x is an integer. Write down the least value of x in the shaded region.

Chapter 16 Estimation and accuracy

Exercise 16A

1 Write the following numbers correct to 3 significant figures.

a	7251	**b**	16.08	**c**	0.039 47
d	6.499	**e**	38 847	**f**	0.050 70
g	8.7979				

2 Write the following numbers correct to 1 significant figure.

a	7251	**b**	16.08	**c**	0.03947
d	6.499	**e**	38 847	**f**	0.050 70
g	8.7979				

3 Work out an estimate for the value of the following expressions by first rounding each number correct to 1 significant figure.

a 6.4×3.8 **b** 12.7×19.8

c 0.78×19.2

d $102.8 \times 0.071 \times 19.7$

e 268.1×0.042 **f** 0.268^2

4 Work out an estimate for the value of the following expressions by first rounding each number correct to 1 significant figure.

a $198 \div 22$ **b** $642 \div 31.4$

c $6.4 \div 0.038$ **d** $728.3 \div 0.096$

e $0.00394 \div 0.0431$

5 Work out an estimate for the value of the following expressions by first rounding each number correct to 1 significant figure.

a $\dfrac{6.4 \times 12.2}{21.1}$ **b** $\dfrac{17.7 \times 31.1}{10.3}$

c $\dfrac{0.58 \times 31.6}{0.63}$ **d** $\dfrac{89.5 \times 41.2}{0.38}$

e $\dfrac{84.2}{2.4 \times 3.8}$ **f** $\dfrac{0.58}{4.2 \times 4.8}$

6 The shop price of an item is its list price plus VAT charged at 17.5%
Work out an estimate for the shop price of the following items

a radio, list price £49

b freezer, list price £209

c computer, list price £899

7 Use a calculator to work out the value of each of the following
i Write down all the figures on your calculator display
ii Write your answer to a sensible degree of accuracy.

a $\dfrac{4.6^2 + 3.8^2}{\sqrt{68}}$ **b** $\dfrac{\sqrt{3.9^2 + 4.7^2}}{6.2}$

c $\dfrac{\sqrt{2.4} + 4.6^2}{6.4 \times 2.4}$ **d** $\dfrac{2.4^3}{\sqrt{68.5}}$

e $\dfrac{5.8^2 - 2.3^2}{\sqrt{6.4} + 2.4}$ **f** $\dfrac{\sqrt{4.88}}{2.33} + \dfrac{3.6^2}{\sqrt{14.1}}$

Exercise 16B

1 The length of a pen is 13 cm correct to the nearest cm.
  i Write down the smallest length it could be.
 ii Write down the longest length it could be.

2 The distance from London to Newcastle is 456 km correct to the nearest km.
  i Write down the shortest distance it could be.
 ii Write down the longest distance it could be.

3 The mass of a block is 60 grams correct to the nearest gram.
  i Write down the smallest mass it could be.
 ii Write down the largest mass it could be.

4 A pencil has a length of 12 cm correct to the nearest cm.
A pencil case has a length of 123 mm correct to the nearest mm.
Explain why the pencil might not fit inside the pencil case.

5 A circle has a radius of 5 cm correct to the nearest cm.
A square has a side of 10 cm correct to the nearest cm.
Explain why the circle might not fit inside the square.

Exercise 16C

1 The mass of a paperweight is 255 grams correct to the nearest gram. The mass of a calculator is 310 grams correct to the nearest gram. Work out the maximum possible total mass of the paperweight and the calculator.

2 James has 2 pieces of wood. The first piece has a length of 324 cm correct to the nearest cm. The second has a length of 184 cm correct to the nearest cm.

  a Work out the minimum possible value of the sum of the lengths of the 2 pieces.

  b Work out the maximum possible value of the difference in lengths of the 2 pieces.

3 A block of wood has a mass of 500 grams correct to the nearest 10 grams. How many coins, each of mass 50 grams, correct to the nearest gram, must be used so that they will definitely have a total mass greater than that of the block of wood?

4 $x = 16$, correct to the nearest whole number.
$y = 14$, correct to the nearest whole number.
  i Work out the lower bounds of
  a $x + y$ **b** $x - y$ **c** $\dfrac{x}{y}$
  ii Work out the upper bounds of
  a $x + y$ **b** $x - y$ **c** $\dfrac{x}{y}$

5 $p = 6.8$, $q = 3.2$ both written correct to 1 decimal place.
  i Work out the lower bounds of
  a $p + q$ **b** $2pq$
  c $2p - 3q$ **d** $\dfrac{p}{q}$
  ii Work out the upper bounds of
  a $p + q$ **b** $2pq$
  c $2p - 3q$ **d** $\dfrac{p}{q}$

6 $m = 6.4$, $n = 3.2$ both written correct to 1 decimal place
  i Work out the lower bounds of
  a $m + n$ **b** $m^2 + n^2$ **c** $m^2 - n^2$
  ii Work out the upper bounds of
  a $m + n$ **b** $m^2 + n^2$ **c** $m^2 - n^2$

7 $f = 2.52$, $g = 4.60$ both written correct to 3 significant figures.
  i Work out the lower bounds of
  a $f + g$ **b** $2f - g$
  c $\dfrac{g}{f}$ **d** $\dfrac{g + f}{f}$
  ii Work out the upper bounds of
  a $f + g$ **b** $2f - g$
  c $\dfrac{g}{f}$ **d** $\dfrac{g + f}{f}$

8 A formula for the distance travelled s metres, when a car accelerates at a constant rate of a metres per second, for a time t seconds is $s = \frac{1}{2}at^2$. $t = 5.6$ correct to the nearest tenth of a second. $a = 3.2$ correct to one decimal place.
Calculate the upper bound of s.

9 $t = \sqrt{\dfrac{2s}{a}}$ is a formula which gives the time taken t seconds, to travel a distance of s metres at a constant acceleration of a metres per second per second.
$s = 120$ correct to 2 significant figures.
$a = 3.2$ correct to 2 significant figures.
Calculate the lower bound of s.

10 The area of a square is 420 cm² correct to 2 significant figures.
  a Calculate the upper bound of the perimeter of the square. Write down all the figures on your calculator
  b Calculate the lower bound of the length of the diagonal of the square. Write down all the figures on your calculator.

11 A car travels a distance d metres in a time T seconds.
$d = 240$ correct to 2 significant figures
$T = 12.1$ correct to 1 decimal place.
  a Calculate the lower bound of the average speed of the car. Write down all the figures on your calculator.
  b Calculate the upper bound of the average speed of the car. Write down all the figures on your calculator.
  c Round off the lower bound and the upper bound until they both agree.

12 The mass of a cube is 65 kg correct to the nearest kg.
The edge of the cube is 21 cm correct to 2 significant figures
  a Calculate the upper bound for the density of the cube. Write down all the figures on your calculator
  b Calculate the lower bound for the density of the cube. Write down all the figures on your calculator
  c Round off the upper bound and lower bound of the density until they both agree.

Chapter 17 Averages and range

Exercise 17A

1 The lists shows the shoe sizes of ten men.

9   11   10   9   10   8   12   7   10   9

  a   Write down the mode.

  b   Find the median.

  c   Work out the mean shoe size.

2 Here are the marks of eight students in a test.

43   35   67   56   76   48   29   81

Work out the mean mark.

3 Here are the number of children in 5 families.

3   1   4   3   2

  a   Work out the mean number of children per family.

  b   The mean number of children of another 5 families is 1.8 Work out the total number of children in these 5 families.

  c   Find the mean number of children of all ten families

4 Daniel plays a computer game. Here are his scores in six games.

260   245   375   275   195   240

  a   Find his median score.

  b   Work out his mean score.

  c   Daniel plays another game. His mean score is now 300. Work out Daniel's score in this game.

5 The mean weight of 6 bananas is 180 grams.

  a   Work out the total weight of the 6 bananas.

One banana is eaten. The mean weight of the remaining 5 bananas is 175 grams.

  b   Work out the weight of the banana which was eaten.

6 The mean annual salary of ten office workers is £18 000 The mean annual salary of these ten office workers and the office manager is £18 500.
Work out the annual salary of the office manager.

Exercise 17B

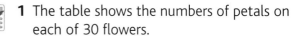

1 The table shows the numbers of petals on each of 30 flowers.

Number of petals	Frequency
8	9
9	8
10	5
11	1
12	7

  a   Write down the modal number of petals.

  b   Work out the median number of petals.

  c   Work out the mean number of petals per flower.

2 The table shows the numbers of passengers in each of 20 taxis arriving at a railway station.

Number of passengers	Frequency
0	3
1	2
2	3
3	5
4	6
5	1

  a   Work out the total number of passengers in these 20 taxis.

  b   Work out the median number of passengers.

  c   Work out the mean number of passengers per taxi.

3 The table gives information about the number of letters received by Mr Lake in the last 20 days.

Number of letters	Frequency
0	3
1	4
2	5
3	8

a Work out the number of letters that Mr Lake received in the last 20 days.

b Find the median number of letters received.

c Work out the mean number of letters received.

4 Here are the scores of 15 golfers in a competition.

68   92   70
72   87   84
76   76   70
75   69   82
83   80   94

a Draw a stem and leaf diagram to show this information.

b Use your stem and leaf diagram to find
  i the range of the scores,
  ii the median score,
  iii the interquartile range of the scores.

Exercise 17C

1 The heights, in metres, of nine students are

1.73   1.56   1.70   1.65   1.68   1.80
1.72   1.64   1.75

For these heights find

a the range   **b** the interquartile range.

2 The stem and leaf diagram shows the number of spelling mistakes made by each student in a class in an essay.

```
0 | 7 9
1 | 1 2 2 3 5 8 8
2 | 0 1 3 5 5 5 7 8 9 9
3 | 0 0 3 5 6
```

KEY   2|3 stands for 23 mistakes

a Find the number of students in the class.

b Write down how many students made more than 25 spelling mistakes.

c Write down how many students made less than 16 spelling mistakes.

Find **d** the range, **e** the median,
    f the interquartile range.

3 Here are the lengths, in cm, of 20 newborn babies.

34   28   19   22   25   38   41   40   27   32
18   21   35   40   37   24   27   33   35   30

Draw a stem and leaf diagram to show this information.   *(1388 March 2006)*

Exercise 17D

1 Twenty people took part in a doughnut eating competition. The number of doughnuts eaten by each person, in a 10-minute period, is shown in the table

Doughnuts eaten	Number of people
1 to 5	0
6 to 10	1
11 to 15	4
16 to 20	6
21 to 25	8
26 to 30	1

a Find the class interval which contains the median.

b Work out an estimate for the mean number of doughnuts eaten.

2 The table gives information about the time taken by 20 students to travel to school.

Time (*t* minutes)	Frequency
$0 < t \leqslant 5$	2
$5 < t \leqslant 10$	8
$10 < t \leqslant 15$	4
$15 < t \leqslant 20$	3
$20 < t \leqslant 25$	3

Work out an estimate for the mean time.
   (1388 March 2006)

3 The table shows information about the number of words in each sentence of an article in a newspaper.

Number of words	Frequency
$0 < t \leqslant 5$	2
$5 < t \leqslant 10$	5
$10 < t \leqslant 15$	12
$15 < t \leqslant 20$	13
$20 < t \leqslant 25$	10
$25 < t \leqslant 30$	8

a Find the class interval which contains the median.

b Work out an estimate for the mean number of words in each sentence.

Exercise 17E

1 The table shows some information about student absences.

Term	Autumn 2003	Spring 2004	Summer 2004
Number of absences	408	543	351

Term	Autumn 2004	Spring 2005	Summer 2005
Number of absences	435	582	372

Work out the three-point moving averages for this information.
The first two have been done for you.
434, 443, ... , ...   *(1388 June 2006)*

2 The table shows the number of people attending the first six home matches of Hawkham United.

	1st	2nd	3rd	4th	5th	6th
Number of people	1240	1354	1306	1390	1378	1420

Find the set of 3-point moving averages for this information.
Comment on the trend of the attendances during this period.

3 A garage sells cars. The table shows the number of cars sold in each half-year period for the last four years.

	Months	Number of cars sold
Year 1	Jan–June	142
	July–Dec	156
Year 2	Jan–June	148
	July–Dec	162
Year 3	Jan–June	142
	July–Dec	152
Year 4	Jan–June	136
	July–Dec	146

a Plot this information on a time series graph.

b Work out the four-point moving averages for this information.

c On your time series graph plot the moving averages.

d Comment on the sales of cars during this period?

Chapter 18 Formulae

Exercise 18A

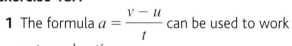

1 The formula $a = \dfrac{v - u}{t}$ can be used to work out acceleration a.
Use this formula to work out the value of a when

a $v = 30, u = 20$ and $t = 2$

b $v = 45, u = 12$ and $t = 3$

2 The formula for the perimeter of a rectangle is $P = 2(l + w)$
Use this formula to find the value of P when

a $l = 5$ and $w = 3$

b $l = 8$ and $w = 2$

c $l = 2.5$ and $w = 3.5$

d $l = 1.2$ and $w = 0.3$

3 Use the formula $A = 3r^2$ to work out the value of A when

a $r = 4$ **b** $r = 10$

4 Use the formula $L = 5\sqrt{\dfrac{p}{3}}$ to work out the value of L when

  a $p = 12$ **b** $p = 75$

5 $S = mx^2 - 5x$
Find the value of S when

  a $x = 2$ and $m = 3$ **b** $x = 1$ and $m = 2$
  c $x = -3$ and $m = 2$ **d** $x - 5$ and $m = 4$

6 The formula $C = \frac{5}{9}(F - 32)$ is used to change degrees Fahrenheit (F) to degrees Celsius (C).
Find the value of C when

  a $F = 68$ **b** $F = 32$
  c $F = 104$ **d** $F = -4$

7 $w = \dfrac{x + 2y}{5}$
Use this formula to work out the value of w,

  a when $x = 4$ and $y = 3$
  b when $x = 7$ and $y = 9$
  c when $x = 8$ and $y = -12$

8 $V = \sqrt{u^2 + 20s}$
Find the value of V when

  a $u = 5$ and $s = 2$
  b $u = 30$ and $s = 14$

9 $T = w(x^2 + y^2)$
Find the value of T when

  a $w = 3$, $x = 2$ and $y = 4$
  b $w = 4$, $x = -5$ and $y = -10$

Exercise 18B

1 A shop sells c CDs at £8 each and d DVDs at £10 each. If T is the total cost of the CDs and DVDs sold, write down, in terms of c and d, a formula for T.

2 A shop sells cups and mugs. Write down an algebraic formula that can be used to work out the total number of cups and mugs that the shop sells each day.
You must define the letters used.

3 a Write down a formula that can be used to work out the perimeter of an regular pentagon.
You must define the letters used.

  b Use your formula to work out the perimeter of an regular pentagon of side 6 cm.

4 Angela rents a villa for her holidays. She pays a fixed charge of £600 for one week plus £50 for each additional day that she stays in the villa.

  a Write down a formula that Angela can use to work out the total cost to rent the villa.
You must define the letters used.

  b Use your formula to work out the cost of renting this villa for
    i 10 days,   **ii** 2 weeks.

5 A box of pencils costs £2 A box of crayons costs £3. Michelle buys p boxes of pencils and c boxes of crayons. The total cost is T pounds.

  a Write a formula for T in terms of p and c.

  b Work out the value of T when,
    i $p = 5$ and $c = 4$ **ii** $p = 12$ and $c = 7$

6 Melissa has two sisters. The mean of her sisters' ages is x years.

  a Write down a formula that can be used to work out the value of x. You must define the letters used.

  b Melissa is 8 years of age. Write down a formula that can be used to work out the mean age of the three sisters.

Exercise 18C

1 Make q the subject of the formula

  a $p = q + 3$ **b** $p = q - 7$

  c $p = 5q$ **d** $p = \dfrac{q}{4}$

  e $p = 2q - 9$ **f** $p = 5 - 3q$

2 Make c the subject of the formula
$f = 3c - 4$

(1388 March 2006)

3 Make b the subject of the formula
$A = \frac{1}{2}bh$.

4 Simple interest is found using the formula
$I = \dfrac{PRT}{100}$. Rearrange the formula to make P the subject.

5 Make x the subject of the formula

  a $y = \dfrac{2x + 3}{5}$ **b** $y = 2(x - 5)$

  c $y = 4(1 - 3x)$ **d** $y = 5x + 2(x + 1)$

  e $y = 3(x - 2) + 2(x + 1)$

6 Make f the subject of the formula
$g = \dfrac{5e + 4f}{3}$.

7 The volume of a cylinder is given by the formula $V = \pi r^2 h$.

  a Make r the subject of the formula.

  b Find the positive value of r when $V = 40$ and $h = 5$

8 a Make t the subject of the formula
$d = 20 - at^2$.

  b Find the values of t when $d = 8$ and $a = 5$.

Exercise 18D

1 Write down whether each of the following is an expression or an identity or an equation or a formula.

  a $L = M + 7$ **b** $A = 2bh$

  c $3t + 2t = 5t$ **d** $y - 3$

  e $5s + 4 = 12$ **f** $x^3 - x^2$

  g $5pq$ **h** $\dfrac{x - 1}{2} = 5$

  i $C = 8B - 5D$

  j $6y + 4 - 2y = 4y + 4$

  k $m^2 + m - 8 = 0$

  l $v = u + at$

  m $c + 3d - 4c = 3d - 3c$

  n $x^2 + 4x - 12$

  o $7(r + 3) = 7r + 21$

  p $6a + 8b = 2(3a + 4b)$

  q $x - \dfrac{2x + 1}{3}$

Exercise 18E

1 Make n the subject of
$2m - 3n = 1 + 4n$

2 Make x the subject of
$xy + 2x = y(1 - 3x)$

3 Rearrange $\dfrac{2x}{x + 3}$ to make x the subject.

4 Make P the subject of
$3 - TP = Px - T$

5 Make m the subject of the formula
$c = \dfrac{1 - 2m}{1 + 3m}$

6 The area of a rectangle of length r cm and width s cm is equal to the area of a circle of radius r.

  a Express r in terms of s and π.

  b Use your answer to part **a** to find the value of r when $s = 5$. Give your answer in terms of π.

7 Make **a** f, **b** u the subject of the formula
$\dfrac{1}{f} = \dfrac{1}{u} + \dfrac{1}{v}$

Chapter 19 Pythagoras' theorem and trigonometry (1)

Exercise 19A

In this exercise, give answers in cm, correct to 3 significant figures where appropriate.

1 Work out the length of the sides marked with letters in these triangles.

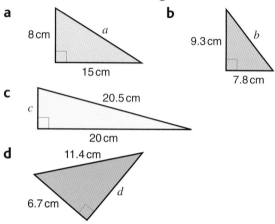

a 8 cm, 15 cm, a

b 9.3 cm, 7.8 cm, b

c 20.5 cm, 20 cm, c

d 11.4 cm, 6.7 cm, d

2 In triangle PQR $QR = 9.3$ cm.
$PQ = 5.7$ cm. Angle $PQR = 90°$

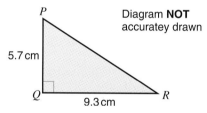

Diagram **NOT** accuratey drawn

P, 5.7 cm, Q, 9.3 cm, R

Calculate the length of PR.

(1388 November 2005)

3 PQR is a right-angled triangle.
Angle $PQR = 90°$.
$QR = 15$ cm.
$PR = 19$ cm.
Work out the length of PQ.

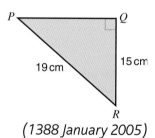

P, Q, 19 cm, 15 cm, R

(1388 January 2005)

4 In triangle ABC,
angle $C = 90°$,
$AB = 14$ cm and
$BC = 9$ cm.
  i Work out the height, AC, of the triangle.
  ii Work out the area of the triangle.

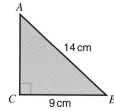

A, 14 cm, C, 9 cm, B

5 Work out the value of x.

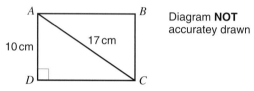

Diagram **NOT** accuratey drawn

7.5 cm, x cm, 7.2 cm

(4400 May 2006)

Exercise 19B

1 $ABCD$ is a rectangle.

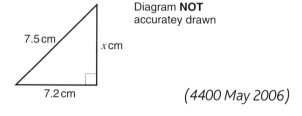

Diagram **NOT** accuratey drawn

A, B, 10 cm, 17 cm, D, C

$AC = 17$ cm. $AD = 10$ cm.
Calculate the length of the side CD.
Give your answer correct to one decimal place.

(1387 November 2004)

2 Triangle ABC is isosceles with $AB = AC$.

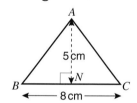

A, 5 cm, N, B, 8 cm, C

The midpoint of BC is the point N.
$AN = 5$ cm and $BC = 8$ cm.
  a Work out the length of AB. Give your answer in cm correct to 1 decimal place.
  b Work out the perimeter of the triangle. Give your answer in cm correct to 1 decimal place.

3

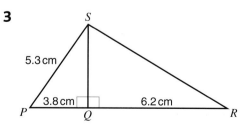

S, 5.3 cm, 3.8 cm, 6.2 cm, P, Q, R

Angle $PQS = 90°$.
Angle $RQS = 90°$.
$PS = 5.3$ cm, $PQ = 3.8$ cm, $QR = 6.2$ cm.

Calculate the length of RS.
Give your answer correct to 3 significant
figures.   *(4400 November 2004)*

4 In the diagram, T is a point on a circle,
centre O.

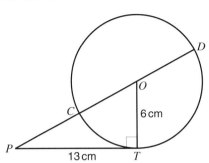

PT is the tangent to the circle at T so that
the angle $OTP = 90°$.
$DOCP$ is a straight line so that CD is a
diameter of the circle.

a Work out the length of OP.
Give your answer in cm correct to
1 decimal place.

b Hence work out the length of
i PC, **ii** PD.
Give your answers in cm correct to
1 decimal place.

5 $PQRS$ is a trapezium with the sides PS and
QR parallel.
PQ is perpendicular to both PS and QR.
$QR = 8$ cm, $RS = 10$ cm,
$PS = 12$ cm.

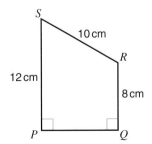

a Work out the length of PQ.
Give your answer in cm correct to
2 decimal places.

b Work out the area of trapezium $PQRS$.
Give your answer to the nearest cm^2.

6 The diagram shows a circle, a chord and two
radii. The radius of the circle is 6.8 cm and
the perpendicular distance of the chord from
the centre of the circle is 5.2 cm.

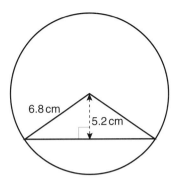

Work out the length of the chord.
Give your answer in cm correct to 2 decimal
places.

Exercise 19C

1 A is the point with coordinates $(6, 3)$.
B is the point with coordinates $(10, 9)$.
M is the midpoint of the line AB.
Work out the coordinates of the point M.

2 A is the point with coordinates $(2, 5)$.
B is the point with coordinates $(8, 13)$.

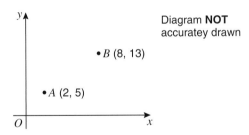

Calculate the length of AB.
                    (1388 April 2006)

3 Two points, A and B, are plotted on a
centimetre grid.
A has coordinates $(2, 1)$ and B has
coordinates $(8, 5)$.

a Work out the coordinates of the
midpoint of the line joining A and B.

b Use Pythagoras' Theorem to work out
the length of AB.
Give your answer correct to 3 significant
figures.
                    (4400 May 2004)

4 The points $P(1, 3)$, $Q(6, 13)$, $R(12, 10)$ and $S(7, 0)$ are the vertices of a quadrilateral $PQRS$.

  a Work out the length of

    i PQ,   **ii** QR,   **iii** RS,   **iv** SP.

    Give each answer correct to 2 decimal places.

  b Explain what your answers to **a** tell you about the quadrilateral $PQRS$.

  c Work out the length of the diagonal PR of $PQRS$.

    Give your answer correct to 2 decimal places.

  d By considering the lengths of the sides of the triangle PQR

    i explain why angle PQR is a right-angle,

    ii explain what this tells you about the quadrilateral $PQRS$.

Exercise 19D

1 Use a calculator to find the value of each of the following.

Give answers correct to 4 decimal places, where necessary.

  a $\sin 50°$ **b** $\sin 81°$ **c** $\cos 45°$

  d $\tan 17°$ **e** $\sin 30°$ **f** $\cos 79°$

  g $\tan 88°$ **h** $\cos 25.7°$ **i** $\tan 63.1°$

  j $\sin 28.3°$ **k** $\tan 9.4°$ **l** $\cos 56.8°$

2 Work out the lengths of the sides marked with letters. Give each answer correct to 3 significant figures.

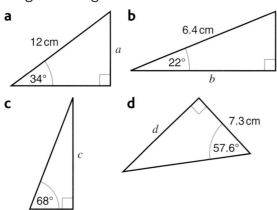

e

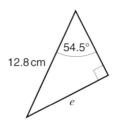

3 Triangle ABC is right-angled at B. In each part calculate the length of BC. Give each answer correct to 3 significant figures.

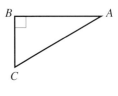

  i $AC = 18.3$ cm, angle $BCA = 73.1°$

  ii $AC = 9.6$ cm, angle $CAB = 42.8°$

  iii $AB = 13.7$ cm, angle $BCA = 38°$

4 The diagram shows a triangle PQR.
$RQ = 12$ cm.
Angle $PQR = 40°$.

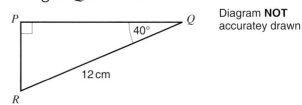

Diagram **NOT** accuratey drawn

Calculate the length of the side PQ.
Give your answer correct to 3 significant figures.

5 Calculate the value of x.

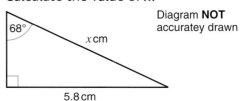

Diagram **NOT** accuratey drawn

Exercise 19E

1 Use a calculator to find the size in degrees of the acute angle x in each of the following. Give answers correct to the nearest $0.1°$ where necessary.

  a $\sin x = 0.7$ **b** $\cos x = 0.23$

  c $\tan x = 0.87$ **d** $\cos x = 0.94$

  e $\sin x = 0.238$ **f** $\tan x = 3.729$

  g $\sin x = \dfrac{7}{11}$ **h** $\tan x = \dfrac{5.9}{8.3}$

2 Work out the size of each of the marked angles. Give each answer correct to 1 decimal place.

a

b

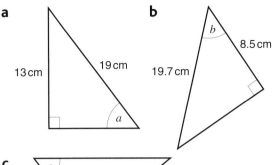

c

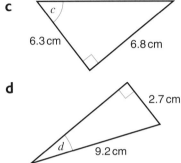

d

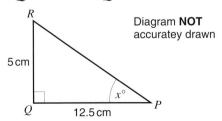

3 Triangle PQR is right-angled at R.
$PR = 4.7$ cm and $PQ = 7.6$ cm.

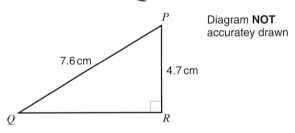

Diagram **NOT** accuratey drawn

Calculate the size of angle PQR.
Give your answer correct to 1 decimal place.
(4400 May 2004)

4 PQR is a triangle.
Angle $PQR = 90°$.
$PQ = 12.5$ cm. $QR = 5$ cm.

Diagram **NOT** accuratey drawn

Calculate the value of x
Give your answer correct to 1 decimal place.
(1387 November 2004)

5 ABC is a right-angled triangle.
Angle $A = 90°$.
$AB = 2.3$ cm. $BC = 5.4$ cm.

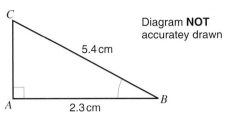

Diagram **NOT** accuratey drawn

Work out the size of angle B.
Give your answer correct to 3 significant figures.
(1388 April 2006)

Exercise 19F

1 $DE = 6$ m. $EG = 10$ m. $FG = 8$ m.
Angle $DEG = 90°$. Angle $EFG = 90°$.

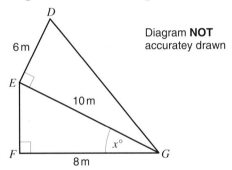

Diagram **NOT** accuratey drawn

a Calculate the length of DG.
Give your answer correct to 3 significant figures.

b Calculate the size of the angle marked $x°$.
Give your answer correct to one decimal place.
(1387 June 2004)

2 ABC is a triangle.
ADC is a straight line with BD perpendicular to AC.
$AB = 7$ cm. $BC = 12$ cm. Angle $BAD = 65°$.

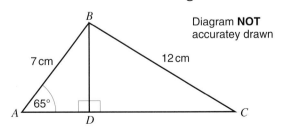

Diagram **NOT** accuratey drawn

Calculate the length of AC.
Give your answer correct to 3 significant figures.
(1388 January 2005)

3 A lighthouse, L, is 3.2 km due West of a port, P.
A ship, S, is 1.9 km due North of the lighthouse, L.

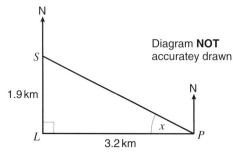

Diagram **NOT** accuratey drawn

a Calculate the size of the angle marked x.
Give your answer correct to 3 significant figures.

b Find the bearing of the port, P, from the ship, S.
Give your answer correct to 3 significant figures.   *(1387 June 2005)*

4 The diagram shows the side view of a rectangular box $ABCD$ on a lorry.

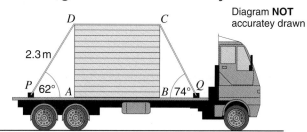

Diagram **NOT** accuratey drawn

The box is held down on the horizontal flat surface of the lorry by a rope.
The rope passes over the box and is tied at two points, P and Q, on the flat surface.
$DP = 2.3$ m.
Angle $APD = 62°$. Angle $BQC = 74°$.
Calculate the length of BQ.
Give your answer correct to 3 significant figures.   *(4400 May 2006)*

5

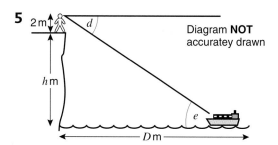

Diagram **NOT** accuratey drawn

The diagram shows a man, of height 2 m standing at the top of a cliff, of height h metres. The man sees a boat on the sea a distance D metres from the foot of the cliff. When $D = 600$, the angle of elevation, e, of the man from the boat is 10°.

a Work out the height, h metres, of the cliff.
Give your answer correct to 3 significant figures.

The boat is moving towards the foot of the cliff and the man remains at the top of the cliff.

b When $D = 400$, work out the angle of depression, d, of the boat from the man.
Give your answer to the nearest degree.

Chapter 20 Ratio and proportion

Exercise 20A

1 Write each ratio in its simplest form.
   a $6:8$ **b** $5:10$ **c** $12:20$
   d $25:40$ **e** $32:24$ **f** $150:200$

2 Write each ratio in its simplest form.
   a £2 : 40p   **b** 30 cm : 1 m
   c 500 m : 3 km **d** 20 mm : 6 cm
   e 2.5 kg : 750 g **f** 30 mm : 0.5 m

3 Write each ratio in its simplest form.
   a $0.2:2.4$ **b** $3.5:4.5$
   c $3.2:4.8$ **d** $0.05:2.5$
   e $0.4:2:3.6$ **f** $0.05:0.3:1.5$

4 Write each ratio in its simplest form.
   a $\frac{1}{3}:\frac{1}{5}$ **b** $\frac{1}{4}:\frac{1}{12}$ **c** $\frac{3}{8}:\frac{1}{2}$
   d $\frac{3}{5}:\frac{1}{10}$ **e** $\frac{2}{3}:\frac{3}{4}:\frac{5}{12}$ **f** $\frac{5}{6}:2:\frac{7}{9}$

5 Write the following ratios in the form $1:n$
   a $2:6$ **b** $5:20$ **c** $2:9$
   d $5:1$ **e** $3:2$ **f** $\frac{5}{9}:\frac{2}{3}$

6 There are 20 sweets in a packet. 12 of the sweets are mints. The rest of the sweets are toffees. Write down the ratio of the number of mints to the number of toffees.
Give your ratio in its simplest form.
(1388 March 2006)

7 A recipe for cake requires
450 grams of flour
and 175 grams of butter.
Write down the ratio of grams of flour to grams of butter.
Give your answer in its simplest form.
(1388 January 2003)

8 In a village there are 300 families and a total of 630 children.
a Write down the ratio of the number of families to the number of children. Give your ratio in its simplest form.
b Write your answer to **a** in the form $1:n$.

9 In a school photograph the ratio of the number of adults to the number of boys to the number of girls is $1:5:6$
a What fraction of the people in the photograph are adults?
b What fraction of the people in the photograph are girls? Give your fraction in its simplest form.

Exercise 20B

1 The ratio to the number of number of red beads to the number of yellow beads in a bag is $1:2$
Work out the number of yellow beads if there are
a 2 red beads　　**b** 5 red beads
c 13 red beads

2 Purple paint is made by mixing blue paint and red paint in the ratio $2:3$
Work out the amount of red paint needed for
a 4 litres of blue paint
b 10 litres of blue paint
c 16 litres of blue paint.

3 In a garden the ratio of the number of daffodils to the number of tulips is $5:3$
a If there are 15 tulips in the garden, work out the number of daffodils.
b If there are 60 daffodils in the garden, work out the number of tulips.

4 In a field the ratio of the number of horses to the number of cows is $1:6$
a If there are 3 horses in the field, work out the number of cows.
b If there are 48 cows in the field, work out the number of horses.

5 On a map, 1 cm represents 4 km. What distance on the map will represent a real distance of
a 12 km　　**b** 32 km　　**c** 10 km?

6 On a map, 1 cm represents 4 km. Work out the real distance between two towns if their distance apart on the map is
a 2 cm　**b** 6 cm　**c** 3.5 cm　**d** 5.3 cm

7 James uses a scale of $1:200$ to make a scale drawing of a building.
a On the scale drawing, the width of the building is 3 cm. What is the real width of the building?
b On the scale drawing, the length of the building is 4.3 cm. What is the real length of the building?

8 The scale of a map is $1:50\,000$
a On the map, the distance between two towns is 3.1 cm. Work out the real distance between the towns. Give your answer in kilometres.
b Work out the distance on the map between two towns if the real distance between the towns is **i** 5 km **ii** 8 km

9 The length of a coach is 15 metres. Jonathan makes a model of the coach. He uses a scale of $1:24$
Work out the length, in centimetres, of the model coach.
(1387 June 2005)

10 The width and length of a rectangle are in the ratio 4 : 7. If the width of the rectangle is 45.6 cm, work out the length of the rectangle.

11 In a business, the ratio of the number of employees to the number of computers is 1 : $\frac{5}{6}$ If there are 402 employees, work out the number of computers.

Exercise 20C

1 a Share £30 in the ratio 1 : 2

b Share £25 in the ratio 3 : 2

2 Colin and David share £20 in the ratio 1 : 4 Work out how much each person gets.

3 A piece of wood is of length 45 cm. The length is divided in the ratio 7 : 2 Work out the length of each part.
(1388 Nov 2005)

4 Alex and Ben were given a total of £240 They shared the money in the ratio 5 : 7 Work out how much money Ben received.
(1388 January 2005)

5 Ken and Susan share £20 in the ratio 1 : 3 Work out how much money each person gets.
(1388 June 2003)

6 Mrs Smith shared £375 between her two children in the ratio 1 : 4 She gave the bigger share to Kylie. Work out how much money she gave to Kylie.

7 Derek, Erica and Fred share £108 in the ratio 3 : 4 : 2 Calculate the amount of money Erica gets.
(1388 January 2003)

8 There are 21 questions in a science test. Each question is on biology or on chemistry or on physics. The numbers of questions on biology, chemistry and physics are in the ratio 4 : 2 : 1

i What fraction of the questions are on chemistry?

ii Work out the number of questions that are on biology.
(1388 March 2006)

Exercise 20D

1 4 sweets cost 20p. Work out the cost of 7 of these sweets.

2 Richard paid 56p for 7 pencils. The cost of each pencil was the same. Work out the cost of 4 of these pencils.
(1388 June 2004)

3 Michael buys 3 files. The total cost of these 3 files is £5.40 Work out the total cost of 7 of these files.
(1387 June 2005)

4 Ruth makes poached peaches. Here is a list of ingredients for making poached peaches for 6 people.

Poached Peaches
Ingredients for 6 people
12 yellow cling peaches
1400 ml water
130 g granulated sugar

Ruth makes poached peaches for 9 people. Work out the amount of each ingredient needed to make poached peaches for 9 people.
(1388 Nov 2005)

5 This is a list of ingredients for making a pear & almond crumble for 4 people.

Ingredients for **4** people

80 g plain flour
60 g ground almonds
90 g soft brown sugar
60 g butter
4 ripe pears

Work out the amount of each ingredient needed to make a pear & almond crumble for **10** people.
(1387 June 2004)

6 Alfie went to Spain.
He changed £400 into euros.
The exchange rate was £1 = 1.45 euros.
Work out the number of euros that Alfie got.
(1388 March 2006)

7 A student bought a pair of sunglasses in the USA.
He paid $35.50.
In England, an identical pair of sunglasses costs £26.99.
The exchange rate is £1 = $1.42.
In which country were the sunglasses cheaper and by how much?
Show all your working. *(1388 June 2004)*

Exercise 20E

1 It takes 3 men 12 days to dig a ditch. Work out how long it will take to dig the ditch if there are

a 2 men   **b** 6 men   **c** 9 men

2 Mrs Smart has enough food to feed 4 cats for 3 days. Work out the number of days that the same amount of food will feed 6 cats.

3 A journey takes 5 hours at an average speed of 80 km/h. If the same journey takes 4 hours work out the new average speed.

4 A quantity of dough makes 24 rolls of 25 g each. If the size of the rolls is increased to 30 g, work out the number of rolls that the same quantity of dough will make.

5 It takes 6 women 9 hours to make enough squares for a blanket.

a How long will it take 18 women to make the same blanket?

b How long will it take 3 women to knit enough square for 2 blankets?

6 A document will fit exactly onto 50 pages if there are 360 words on a page. If the number of words on each page is increased to 400, how many *fewer* pages will there be in the document?

7 In an exam hall there are 10 rows of desks with 24 desks in each row. If the desks are rearranged into rows of 16 desks, work out the number of rows.

Chapter 21 Processing, representing and interpreting data

Exercise 21A

1 Mr Potts records the number of customers that come into his shop over a period of 30 days.

37 32 48 53 32 47 42 45 33 49
52 32 34 31 38 40 52 36 37 33
34 39 43 52 50 46 44 37 38 32

a Copy and complete the grouped frequency table.

Number of customers	Tally	Frequency
31–35		
36–40		
41–45		
46–50		
51–55		

b Write down the modal class interval.

c Draw a frequency diagram for this information.

2 The histogram shows information about the heights of the girls in a class.

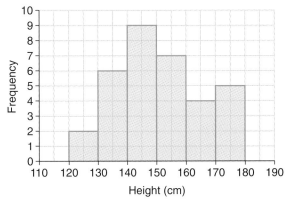

a Write down the modal class interval.

b Work out the total number of girls.

3 The grouped frequency table gives information about the amount of time patients had to wait in a doctor's surgery.

Time (*m* minutes)	Frequency
$0 \leqslant m < 5$	2
$5 \leqslant m < 10$	5
$10 \leqslant m < 15$	8
$15 \leqslant m < 20$	6
$20 \leqslant m < 25$	3
$25 \leqslant m < 30$	2

a Draw a histogram to show this information.

b Draw a frequency polygon to show this information.

4 32 students took an English test.
There were 25 questions in the test.
The grouped frequency table gives information about the number of questions the students answered.

Number of test questions answered	Frequency
1–5	1
6–10	3
11–15	9
15–20	8
21–25	11

a Write down the modal class interval.

b Write down the class interval which contains the median.

c On graph paper, draw a frequency polygon to show the information in the table.

(1385 November 2002)

Exercise 21B

1 The following cumulative frequency table gives information about the heights, in centimetres, of some plants.

a On graph paper, draw a cumulative frequency graph for this data.

Height (*h* cm)	Cumulative frequency
$0 < h \leqslant 5$	2
$0 < h \leqslant 10$	8
$0 < h \leqslant 15$	20
$0 < h \leqslant 20$	32
$0 < h \leqslant 25$	37
$0 < h \leqslant 30$	40

b Use your graph to find an estimate for the number of plants that had a height **i** less then 17 cm **ii** more than 17 cm

2 The grouped frequency table gives information about the time taken by 60 students to complete a 2 hour mathematics exam.

Time (*t* minutes)	Frequency
$0 < t \leqslant 20$	1
$20 < t \leqslant 40$	5
$40 < t \leqslant 60$	10
$60 < t \leqslant 80$	16
$80 < t \leqslant 100$	25
$100 < t \leqslant 120$	3

a Copy and complete the cumulative frequency table.

Time (*t* minutes)	Cumulative frequency
$0 < t \leqslant 20$	
$0 < t \leqslant 40$	
$0 < t \leqslant 60$	
$0 < t \leqslant 80$	
$0 < t \leqslant 100$	
$0 < t \leqslant 120$	

b Use your table to draw a cumulative frequency graph.

c Use your graph to find an estimate for the number of students who completed the paper in less than $1\frac{1}{2}$ hours.

3 90 students took a test.
The cumulative frequency graph gives information about their marks.

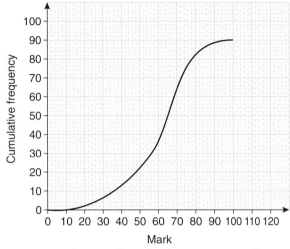

a Use the graph to find an estimate for the number of students whose mark is *less* than 40.

b Use the graph to work out an estimate for the number of students whose mark is *more* than 70.

c 70 students passed the test. Work out the pass mark for the test.

Exercise 21C

1 The cumulative frequency diagram below gives information about the prices of 120 houses.

a Find an estimate for the number of houses with prices less than £130 000.

b Work out an estimate for the interquartile range of the prices of the 120 houses.

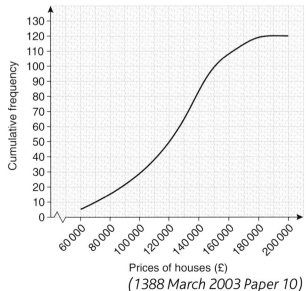

Prices of houses (£)

(1388 March 2003 Paper 10)

2 At a supermarket, members of staff recorded the lengths of time that 80 customers had to wait in the queues at the check-outs.

The waiting times are grouped in the frequency table below.

Waiting time (*t* seconds)	Frequency
$0 < t \leqslant 50$	4
$50 < t \leqslant 100$	7
$100 < t \leqslant 150$	10
$150 < t \leqslant 200$	16
$200 < t \leqslant 250$	30
$250 < t \leqslant 300$	13

a Copy and complete the cumulative frequency table below.

Waiting time (*t* seconds)	Cumulative frequency
$0 < t \leqslant 50$	
$0 < t \leqslant 100$	
$0 < t \leqslant 150$	
$0 < t \leqslant 200$	
$0 < t \leqslant 250$	
$0 < t \leqslant 300$	

b On graph paper, draw a cumulative frequency graph for this data.

c Use your graph to work out an estimate for
  i the median waiting time,
  ii the number of these customers who had to wait for more than 3 minutes.
(1385 November 2000)

3 The table gives information about the weights, in kilograms, of 100 pigs.

Weight of pigs (*w* kg)	Frequency
$65 < w \leqslant 70$	4
$70 < w \leqslant 75$	10
$75 < w \leqslant 80$	34
$80 < w \leqslant 85$	32
$85 < w \leqslant 90$	16
$90 < w \leqslant 95$	4

a Work out the class interval which contains the median.

b Copy and complete the table below to show the cumulative frequency for this data.

Weight of pigs (w kg)	Cumulative frequency
$65 < w \leq 70$	4
$65 < w \leq 75$	
$65 < w \leq 80$	
$65 < w \leq 85$	
$65 < w \leq 90$	
$65 < w \leq 95$	100

c On graph paper, draw a cumulative frequency graph for the data.

d Use your graph to work out an estimate for
  i the interquartile range,
  ii the number of pigs which weigh **more** than 87.5 kg. *(1385 May 2002)*

Exercise 21D

1 A company has 80 employees.

The age of the youngest employee is 22 years.
The age of the oldest employee is 57 years.

The lower quartile age is 31 years.
The median age is 35 years.
The upper quartile age is 47 years.

Draw a box plot to show information about the ages of the employees.

2 The number of telephone calls made to an office for each of 15 days are listed in order.

14   15   15   17   21   23   23   26
31   34   34   37   42   45   46

a Find **i** the median **ii** the interquartile range.

b Draw a box plot for this data.

3 The box plot shows information about the marks scored in a test by some students.

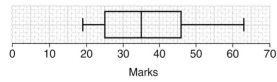

Marks

a Write down the median mark.

b Work out the range of the marks.

4 200 students took a test.
The cumulative frequency graph gives information about their marks.

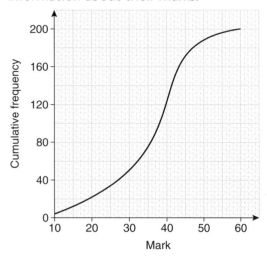

Mark

The lowest mark scored in the test was 10.
The highest mark scored in the test was 60.

Use this information and the cumulative frequency graph to draw a box plot showing information about the students' marks.

(1388 January 2004)

Exercise 21E

1 The histogram gives information about the times, in minutes, 135 students spent on the Internet last night.

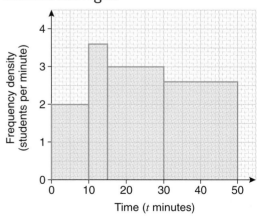

Time (t minutes)

Use the histogram to complete the table.

Time (t minutes)	Frequency
$0 < t \leq 10$	
$10 < t \leq 15$	
$15 < t \leq 30$	
$30 < t \leq 50$	

(1387 June 2004)

2 The table gives information about the heights, in centimetres, of some 15 year old students.

Height (*h* cm)	$145 < h \leqslant 155$	$155 < h \leqslant 175$	$175 < h \leqslant 190$
Frequency	10	80	24

Use the table to draw a histogram.

(1388 January 2003)

3 A teacher asked some year 10 students how long they spent doing homework each night. The histogram was drawn from this information.

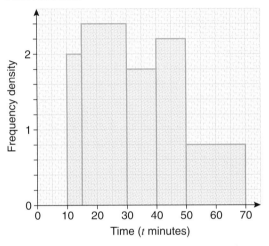

Use the histogram to copy and complete the table.

Time (*t* minutes)	Frequency
$10 \leqslant t < 15$	10
$15 \leqslant t < 30$	
$30 \leqslant t < 40$	
$40 \leqslant t < 50$	
$50 \leqslant t < 70$	

(1388 March 2003)

4 The table gives information about the times, in hours, some students took to complete a piece of coursework. Use this information to draw a histogram.

Time (*t*) in hours	Frequency
$0 < t \leqslant 20$	30
$20 < t \leqslant 30$	40
$30 < t \leqslant 55$	25

(1388 March 2005)

Chapter 22 Three-dimensional shapes

Exercise 22A

1 Copy these diagrams and draw in one plane of symmetry for each of these shapes.

i

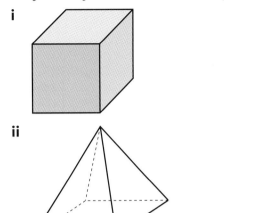

ii

(1385 June 2000)

2 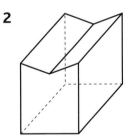 The diagram represents a prism.
Copy the diagram and draw in one plane of symmetry of the prism.

(1384 June 1997)

3 The diagram shows a sketch of a solid. Each end of the solid is an isosceles triangle.

a Write down the name of the solid.

b For the solid, write down
i the number of faces,
ii the number of edges,
iii the number of vertices.

c Copy the diagram and draw one of the solid's planes of symmetry.

d How many planes of symmetry does the solid have?

(1387 June 2002)

Exercise 22B

1 On centimetre squared paper, draw the plan, front elevation and side elevation for this prism.

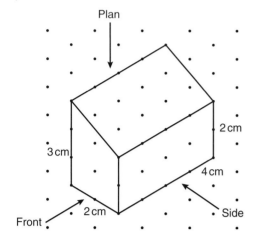

Plan

2 cm

3 cm

4 cm

2 cm

Front

Side

2 Draw a sketch of the plan, front elevation and side elevation of this solid.

3 Here are the plan, front elevation and side elevation of a 3-D shape.
On isometric paper, draw the prism.

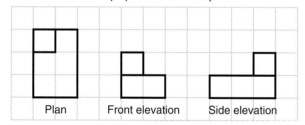

Plan   Front elevation   Side elevation

Exercise 22C

1

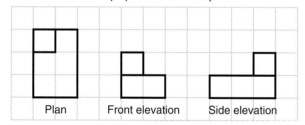

a Write down the coordinates of S, T and U.

b Write down the coordinates of the midpoint of ST.

c Write down the coordinates of the midpoint of TU.

2 The coordinates of five of the corners of a cube are $(0, 0, 0)$, $(2, 0, 0)$, $(2, 2, 0)$, $(0, 2, 0)$ and $(0, 0, 2)$. Find the coordinates of the other three corners.

3 The coordinates of five of the corners of a cuboid are $(1, 3, -2)$, $(1, 3, 1)$, $(1, -2, 1)$, $(1, -2, -2)$ and $(-4, 3, -2)$. Find the coordinates of the other three corners.

4 A is the point $(2, 2, 0)$. B is the point $(0, 0, 6)$. Find the coordinates of the midpoint of AB.

Exercise 22D

1 Work out the volume of a cuboid which is 10 cm by 8 cm by 7 cm.

2 Work out the volume of a cuboid which is 20 m by 8 m by 5 m.

3 Work out the volume of a cuboid which is 12 mm by 10 mm by 6 mm.

4 Work out the volume, in cm³, of a cuboid which is 3 m by 6 cm by 4 cm.

5 Work out the volume, in m³, of a cuboid which is 4 m by 80 cm by 5 cm.

6 The volume of a cuboid is 120 cm³. Its length is 8 cm and its width is 5 cm. Work out its height.

7 The volume of a cuboid is 840 m³. Its length is 20 m and its height is 6 m. Work out its width.

8

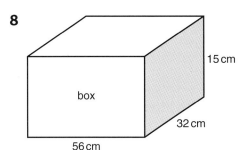

15 cm

box

32 cm

56 cm

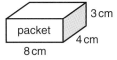

3 cm

packet

4 cm

8 cm

A packet is a cuboid
which is 8 cm by 4 cm
by 3 cm.
A box is a cuboid which
is 56 cm by 32 cm by 15 cm.
Work out how many packets will fit exactly
into the box.

9 A rectangular box measures 10 cm by 8 cm
by 2.5 cm.
A container is a cuboid which measures
80 cm by 40 cm by 30 cm.
Work out the greatest number of boxes
which can be packed in the container.

Exercise 22E

1 The diagram shows a prism with a right-
angled triangle as its cross-section.

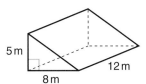

5 m

8 m

12 m

Work out the volume of the prism.

2 The diagram shows a prism.

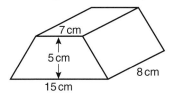

7 cm

5 cm

8 cm

15 cm

The cross-section of the prism is a trapezium.
The lengths of the parallel sides of the
trapezium are 15 cm and 7 cm.
The distance between the parallel sides of
the trapezium is 5 cm.
The length of the prism is 8 cm.
Work out the volume of the prism.

3 The radius of the end of a cylinder is 4 mm.
Its length is 9 mm.

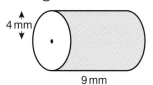

4 mm

9 mm

Work out the volume of the cylinder.
Give your answer correct to the nearest mm^3.

4 The diagram shows the cross-section of a
prism.

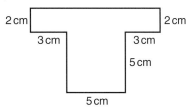

2 cm

2 cm

3 cm

3 cm

5 cm

5 cm

All the corners are right angles.
The length of the prism is 10 cm.
Work out the volume of the prism.

5 The cross-section of a prism is a semicircle
of diameter 3.6 cm.

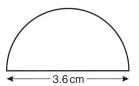

3.6 cm

The length of the prism is 9 cm.
Work out the volume of the prism.
Give your answer correct to 1 decimal place.

Exercise 22F

Where necessary, give answers correct to
3 significant figures.
If your calculator does not have a π button take
the value of π to be 3.142, unless the question
instructs otherwise.

1 The base of a pyramid is a 12 cm by 10 cm
rectangle. Its height is 14 cm.
Calculate its volume.

2 The radius of the base of a cone is 5 cm and
its height is 9 cm.
Calculate the volume of the cone.

3 The radius of the base of a cone is 6 cm and its height is 10 cm.
Find the volume of the cone. Give your answer as a multiple of π.

4 The radius of a sphere is 1.9 cm. Calculate its volume.

5 The radius of a sphere is 9 cm. Find its volume. Give your answer as a multiple of π.

6 A pyramid has a height of 18 cm and a volume of 120 cm^3. Calculate the area of its base.

7 A cone has a volume of 10 m^3.
The vertical height of the cone is 1.5 m.
Calculate the radius of the base of the cone.
Give your answer correct to 3 significant figures.   *(1387 June 2004)*

8 The volume of a sphere is 70 cm^3.
Calculate its radius.

9

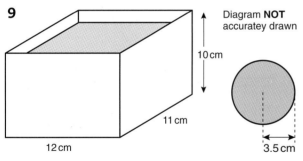

Diagram **NOT** accuratey drawn

10 cm

11 cm

12 cm

3.5 cm

A rectangular container is 12 cm long, 11 cm wide and 10 cm high.
The container is filled with water to a depth of 8 cm.
A metal sphere of radius 3.5 cm is placed in the water.
It sinks to the bottom.
Calculate the rise in the water level.
Give your answer correct to 3 significant figures.   *(1388 November 2005)*

Exercise 22G

1 Change

   a 6.5 cm^3 to mm^3 **b** 800 mm^3 to cm^3.

2 Change

   a 8 m^3 to cm^3 **b** 17 500 000 cm^3 to m^3.

3 Change

   a 5000 cm^3 to litres   **b** 4.9 litres to cm^3.

4 A bottle holds 0.7 litre of wine.
Change 0.7 litre to cm^3.

5 The length of each side of a cube is 50 cm.
Work out the volume of the cube

   a in cm^3 **b** in m^3.

6 A rectangular wall is 4 m wide and 2.5 m high.
One litre of paint is needed to cover the wall.
Work out the thickness of the coat of paint.
Give your answer in millimetres.

Exercise 22H

1 Work out the surface area of

   a a cube of side 10 cm

   b a cuboid which is 5 cm by 4 cm by 8 cm.

2

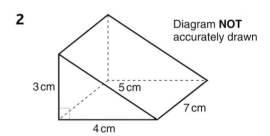

Diagram **NOT** accurately drawn

3 cm   5 cm

7 cm

4 cm

The diagram shows a prism.
The cross-section of the prism is a right-angled triangle.
The lengths of the sides of the triangle are 3 cm, 4 cm and 5 cm.
The length of the prism is 7 cm.

   a Work out the volume of the prism.

   b Work out the total surface area of the prism.
                         (4400 November 2004)

3 The radius of the end
of a solid cylinder
is 4 cm.
Its length is 9 cm.
Work out the total
surface area of the cylinder.
Give your answer correct to the nearest cm².

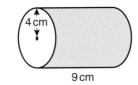

4 cm

9 cm

Exercise 22I

Where necessary, give answers correct to
3 significant figures.
If your calculator does not have a π button take
the value of π to be 3.142, unless the question
instructs otherwise.

1 Calculate the surface area of a sphere of
radius 8.1 cm.

2 The radius of the base of a cone is 5.2 cm
and its slant height is 8.7 cm.
Calculate its curved surface area.

3 The radius of the base of a cone is 1.3 cm
and its slant height is 1.9 cm.
Calculate its total surface area.

The diagrams show the nets of three cones.
Calculate the total surface area of each cone.

4

72°

8.5 cm

1.7 cm

5

135°   5.6 cm

6

5.4 cm

160°

7 The radius of a sphere is 5 cm. Find its
surface area, giving the answer as a multiple
of π.

8 The surface area of a sphere is 100 cm².
Calculate its radius.

Chapter 23 Graphs (2)

Exercise 23A

1 a Copy and complete this table of values
for $y = x^2 + 1$

x	−2	−1	0	1	2	3
y	5		1			10

b Draw the graph of $y = x^2 + 1$

c Write down the coordinates of the
minimum point.

2 a Copy and complete this table of values
for $y = 5 - 2x^2$

x	−3	−2	−1	0	1	2	3
y	−13		3			−3	

b Draw the graph of $y = 5 - 2x^2$

c Write down an estimate for the
coordinates of the points where the
graph crosses the x-axis

3 a Complete this table of values for
$y = x - x^2$

x	−2	−1	0	1	2	3
y		−2	0		−2	

b Draw the graph of $y = x - x^2$ from
$x = -2$ to $x = 3$

c Write down the values of x where the
graph crosses the x-axis.

d Write down the equation of the line of
symmetry of your graph.

e Use your graph to find an estimate for
the maximum value of y.

4 a Complete the table of values for
$y = x^2 - 3x + 1$

x	−2	−1	0	1	2	3	4
y	11		1	−1			5

b Draw the graph of $y = x^2 - 3x + 1$ from
$x = -2$ to $x = 4$ (1387 June 2006)

Exercise 23B

1 a Draw the graph of $y = x^2 - 4$ from $x = -3$ to $x = 3$

b Use the graph of $y = x^2 - 4$ to find the solutions to the equation $x^2 - 4 = 0$

c i On the same axes, draw the graph of $y = 2$

ii Write down the values of the x-coordinates of the points where the two graphs cross.

iii Write down the equation solved by your answers to part **ii**

2 a Draw the graph of $y = x^2 + 2x - 3$ from $x = -4$ to $x = 2$

b Use the graph of $y = x^2 + 2x - 3$ to find the solutions of the equation $x^2 + 2x - 3 = 0$

3 a Draw the graph of $y = 2x^2 - 3x - 2$ from $x = -2$ to $x = 4$

b Use the graph to find

i an estimate for the minimum value of y.

ii solutions of the equation $2x^2 - 3x - 2 = 0$

4 a Draw the graph of $y = 6 - 2x^2$ taking values of x from -3 to 3

b Use the graph to find

i the maximum value of y.

ii estimates of the solutions of the equation $6 - 2x^2 = 0$

iii estimates of the solutions of the equation $10 - 2x^2 = 0$

Exercise 23C

1 a Draw the graph of $y = x^2$ from $x = -3$ to $x = 3$

b On the same axes, draw the graph of $y = 2 - x$

c i Write down the x-coordinates of the two points where the graphs cross.

ii Hence find the quadratic equation whose solutions are these values.

2 By drawing suitable straight lines on the graph of $y = x^2$, estimate the solutions of the equation

a $x^2 = x + 2$ **b** $x^2 - 3x - 4 = 0$

c $2.5x - 1 - x^2 = 0$

3 a Draw the graph of $y = x^2 + 2x - 4$ from $x = -4$ to $x = 2$

b By drawing suitable straight lines on the graph, estimate the solutions of the equation

i $x^2 + 2x - 4 = x - 1$

ii $x^2 + 3x - 2 = 0$

4 a Draw the graph of $y = 4 - x^2$ from $x = -3$ to $x = 3$

b By drawing suitable straight lines on the graph, find estimates of the solutions of the equation

i $4 - x^2 = x + 1$

ii $2 - 2x - x^2 = 0$

iii $x^2 - 2x - 3 = 0$

Chapter 24 Probability

Exercise 24A

1 A fair dice has 3 red faces, 2 blue faces and 1 green face. The dice is thrown once. Write down the probability that the dice will land on

a red,   **b** blue,   **c** green.

2 Michelle has a bag of 10 counters. 3 of the counters are black, 2 are white and the others are yellow. Michelle chooses a counter at random from the bag. Find the probability that Michelle will choose

a a black counter,   **b** a white counter,

c a yellow counter.

3 Jo chooses a letter at random from the letters in the word MISSISSIPPI. Find the probability that she will choose

a a letter S,   **b** a consonant.

4 Uzma has 12 earrings in a jewellery box. 8 earrings are gold and the rest are silver. Uzma chooses an earring at random from the jewellery box. What is the probability that this earring is

   a gold,   **b** silver,   **c** platinum?

5 Dan has 6 cards numbered 2, 4, 5, 8, 9 and 10. Dan chooses one of the cards at random. What is the probability that the card will show

   a an odd number,

   b an odd number less than 5,

   c an even number,

   d a factor of 10,

   e an even number bigger than 5,

   f a factor of 12?

Exercise 24B

1 A 3-sided spinner, with sides numbered 1, 2 and 3, is spun and a dice is rolled at the same time. Draw the sample space showing all possible outcomes.

2 There are three beads in a bag. One bead is red, one bead is white and one bead is yellow. There are also three beads in a box. One bead is green, one bead is pink and one bead is blue. Without looking, Saskia takes, at random, one bead from the bag and one bead from the box. One possible outcome for the colours of the two beads taken is (red, green). List all the possible outcomes.
(1388 March 2006)

3 A fair blue dice and a fair red dice are thrown at the same time.

   a Find the probability that the sum of the numbers on the two dice will be
     i equal to 7   **ii** an even number

   b Find the probability that the number on the blue dice will be greater than the number on the red dice.

4 Zeb has three cards, each with one letter, Z, E, B on. He picks a card at random, notes the letter, then replaces it. He does this one more time.

   a Draw a sample space showing all possible outcomes. One possible outcome is (Z, E).

   b Find the probability that the two cards chosen will show the same letter.

   c Find the probability that at least one of the cards will be a Z.

Exercise 24C

1 The probability that Daniel will be the player of the year is 0.68 Work out the probability that Daniel will not be the player of the year.

2 The probability that it will rain in London tomorrow is $\frac{3}{5}$. Work out the probability that it will not rain in London tomorrow.

3 Each day, Anthony travels to work. He can be on time or early or late. The probability that he will be on time is 0.02 The probability that he will be early is 0.79 Work out the probability that he will be late.
(1388 March 2006)

4 A school snack bar offers a choice of four snacks. The four snacks are burgers, pizza, pasta and salad. Students can choose one of these four snacks. The table shows the probability that a student will choose burger or pizza or salad.

Snack	burger	pizza	pasta	salad
Probability	0.35	0.15		0.2

Work out the probability that the student

   a did not choose salad,

   b chose pasta.   *(1387 November 2005)*

5 A biased dice has coloured faces. The faces are coloured blue or red or yellow or green. The probability that the dice will land on each of the colours red or yellow is given in the table. The probability that the dice will land on blue is double the probability that the dice will land on green.

Colour	blue	red	yellow	green
Probability	$2x$	0.22	0.3	x

Work out the value of x.

Exercise 24D

1 A dice is biased. The dice is thrown 400 times. It lands 150 times on the number 4.

   a Write down the relative frequency of the dice landing on the number 4.

   b The dice is to be thrown again. Estimate the probability that the coin will land on 4.

2 The probability of a student choosing a pizza at a snack bar is 0.35.
400 students used the snack bar on Friday. Work out an estimate for the number of students who chose a pizza.
(1387 November 2005)

3 A box contains 50 counters. There are 23 white counters, 19 black counters and 8 yellow counters. Piero takes at random a single counter from the box. Work out the probability that he takes a white counter or a yellow counter
(1388 November 2005)

4 A bag contains 10 coloured sweets. Each sweet is green or orange or lemon.
Viv chooses a sweet at random from the bag and then replaces it. She does this 400 times. The table shows the numbers of each coloured sweet chosen.

Green	Orange	Lemon
120	202	78

   a Estimate the number of Green sweets in the bag.

   b Estimate the number of Orange sweets in the bag.

   c Estimate the number of Lemon sweets in the bag.

Exercise 24E

1 A bag contains 6 blue counters and 4 white counters. A counter is chosen at random and then returned to the bag. A second counter is then chosen at random.

Work out the probability that

   a both counters will be blue,

   b both counters will be white,

   c one counter will be blue and one counter will be white

2 A coin and a dice are thrown.
The probability that the coin will land on heads is 0.35 The probability that the dice will land on an even number is 0.4.

   a Write down the probability that the dice will not land on an even number.

   b Find the probability that the coin will land on heads and that the dice will land on an even number.

   c Find the probability that the coin will **not** land on heads and that the dice will **not** land on an even number.

3 A box contains 3 bars of milk chocolate, 5 bars of plain chocolate and 4 bars of white chocolate. A bar of chocolate is chosen at random and then returned to the box.
A second bar of chocolate is then chosen at random.
Work out the probability that the bars of chocolate chosen will be

   a both milk chocolate,

   b both plain chocolate,

   c one plain chocolate and one white chocolate.

4 Steven and Frank each take a penalty in a penalty shoot out.
They each have one attempt.
The events are independent.
The probability that Steven will score is $\frac{3}{5}$.
The probability that Frank will score is $\frac{1}{2}$.

   a Find the probability that both Steven and Frank will score.

   b Find the probability that just one of them will score.

   c Find the probability that neither of them will score.

Exercise 24F

1 Ann plays a game of tennis and a game of squash. The probability that Ann will win her game of tennis is 0.8.
The probability that Ann will win her game of squash is 0.7.
The two events are independent.

a Complete the probability tree diagram.

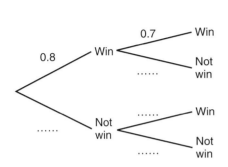

b Work out the probability that Ann will win both games.

c Work out the probability that Ann will win just one game.

2 Alec and Josh sit a Science test.
The probability Alec will pass the test is 0.85.
The probability Josh will pass the test is 0.6.

a Draw a probability tree diagram to show this information

b Find the probability that
  i both students will pass the test,
  ii just Josh will pass the test,
  iii neither will pass the test.

3 The probability that a dice will show a six when thrown is 0.2.
Jo throws the dice twice and records her results.

a Draw a probability tree diagram.

b Use your diagram to work out the probability that the dice will show
  i a six on both throws
  ii a six on **exactly** one throw
  iii a six on at least one throw.

4 A bag contains 12 wooden bricks. 6 bricks are red, 4 bricks are green and 2 bricks are

yellow. Melissa takes a brick at random from the box. She returns the brick to the box and then takes at random a second brick.

a Draw a probability tree diagram.

b Use your diagram to find the probability that
  i both bricks will be of the same colour.
  ii at least one brick will be red.

Exercise 24G

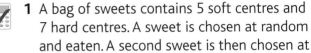

1 A bag of sweets contains 5 soft centres and 7 hard centres. A sweet is chosen at random and eaten. A second sweet is then chosen at random. Work out the probability that for the two sweets chosen

a both will be soft centres,

b the first will be a soft centre and the second will be a hard centre,

c at least one soft centre will be chosen.

2 There are 5 red counters, 3 blue counters and 1 green counter in a bag. A counter is taken at random and **not** replaced. A second counter is then taken at random.

a Find the probability that both counters taken will be
  i red **ii** the same colour

b Find the probability that **exactly** one of the counters will be blue.

3 The probability that it will rain on Monday is 0.36. If it does rain on Monday the probability that it will rain on Tuesday is 0.7. If it does not rain on Monday the probability that it will rain on Tuesday is 0.45.

a Draw a probability tree diagram.

b Using your diagram, find the probability that
  i it will rain on Monday and Tuesday
  ii it will rain on just one day.

4 Robert has 12 notes in his wallet, six £5 notes, five £10 notes and one £20 pound note. He takes two notes at random from his wallet. Work out the probability that two notes each have the same value.

Chapter 25 Further graphs and trial and improvement

Exercise 25A

1 i Do the graphs of the following equations cross the y-axis?

ii If so, find the coordinates of the point where they cross the y-axis.

a $y = 3 - x^3$ **b** $y = 3^x$

c $y = \dfrac{3}{x}$

2 a Complete the table of values for
$y = x^3 - 2x^2 + 5$

x	−2	−1	0	1	2	3
y			5			14

b Draw the graph of $y = x^3 - 2x^2 + 5$ for $-2 \leqslant x \leqslant 3$

3 Draw the graph of $y = \dfrac{8}{1 - x}$ for values of x between -3 and 5.

4 The points with coordinates $(2, 6)$ and $(3, q)$ lie on the graph $y = \dfrac{k}{x}$, where k and q are constants.
Calculate the values of k and q.

5 The points with coordinates $(1, 15)$ and $(3, 375)$ lie on the graph with equation $y = pq^x$, where p and q are constants.
Calculate the values of p and q.

Exercise 25B

1 Tariq and Yousef have been asked to find the solution, correct to one decimal place, of the equation $x^3 + 2x = 56$.

a Work out the value of $x^3 + 2x$ when $x = 3.65$.
Tariq says 3.6 is the solution. Yousef says 3.7 is the solution.

b Use your answer to part **a** to decide whether Tariq or Yousef is correct. You must give a reason.
(1388 March 2006)

2 The equation $x^3 - x = 20$ has a solution between 2 and 3. Use a trial and improvement method to find this solution. Give your answer correct to one decimal place. You must show **ALL** your working.
(1388 January 2004)

3 The equation $x^3 + 10x = 21$ has a solution between 1 and 2. Use a trial and improvement method to find this solution. Give your answer correct to one decimal place. You must show **ALL** your working.
(1387 November 2005)

4 The equation $x^3 + 2x = 65$ has a solution between 3 and 4. Use a trial and improvement method to find this solution. Give your solution correct to one decimal place. You must show **ALL** your working.
(1388 November 2005)

5 Use a trial and improvement method to find the solution of the equation $\dfrac{1}{x} = x^2 + 1$ that lies between 0 and 1. Give your answer correct to one decimal place.

Chapter 26 Indices, standard form and surds

Exercise 26A

1 Find the values of

a 3^{-1} **b** 2^{-2} **c** 4^{-2} **d** 10^{-1}

e 5^0 **f** 0.4^{-1} **g** $\left(\frac{1}{4}\right)^{-1}$ **h** $\left(\frac{3}{5}\right)^{-2}$

2 Simplify

a $4^3 \times 4^{-1}$ **b** $5^2 \times 5^{-5}$

c $2^{-3} \times 2^{-4}$ **d** $3^{-3} \times 3$

e $(6^{-1})^2$

3 Simplify

a $5^{-3} \div 5^{-1}$ **b** $4^{-4} \div 4^{-2}$

c $3^2 \div 3^{-3}$ **d** $4^{-2} \div 4^3$

e $6^{-2} \div 6$

4 Simplify

a $\dfrac{2^5 \times 2^{-2}}{2^2}$ **b** $\dfrac{3^{-4} \times 3^{-3}}{3^{-1}}$

c $\dfrac{5^{-2} \times 5^3}{5^4}$ **d** $\dfrac{3^{-4} \times 3}{3^{-2}}$

5 Simplify

a $\dfrac{3^6}{3^8 \times 3^{-1}}$ **b** $\dfrac{4^5}{4^{-2} \times 4^{-3}}$

c $\dfrac{5^{-3}}{5^2 \times 5}$ **d** $\dfrac{2^8 \times 2^{-5}}{2^6 \times 2^{-3}}$

6 Find the value of n.

a $3^n = \dfrac{3}{3^3}$ **b** $2^{-2} = \dfrac{2^n}{2}$

c $4 \times 4^n = \dfrac{4^2}{4^4}$ **d** $\dfrac{6^n}{6} = 6^{-4}$

e $2^2 \times 2^n = \dfrac{2^5}{2^4}$

7 Use a calculator to work out

a 2.5^{-1} **b** 1.25^{-2} **c** 0.2^{-5}

8 Use a calculator to work out the following. Give your answer correct to 3 significant figures.

a 2.4^{-1} **b** 0.45^{-2} **c** 3.5^{-3}

Exercise 26B

1 Write the following in standard form.

a 80 000 **b** 26 000 **c** 537.25

d 48 **e** 6

2 Write the following as ordinary numbers.

a 4×10^3 **b** 2.4×10^4

c 1.03×10^2 **d** 6×10^2

e 4.6×10^0

3 Write the following in standard form.

a 40×10^6 **b** 36×10^4

c 320×10^4 **d** 200×10^7

e 0.3×10^6 **f** 0.082×10^5

4 Work out the following. Give your answers in standard form.

a $(3 \times 10^8) \times (2 \times 10^2)$

b $(8 \times 10^6) \div (2 \times 10^3)$

c $(3 \times 10^4) \times (8 \times 10^6)$

d $(5 \times 10^{14}) \div (2 \times 10^8)$

e $(2 \times 10^7) \div (8 \times 10^3)$

5 Work out the following. Give your answers in standard form.

a $(4 \times 10^8) \div (8 \times 10^3)$

b $(8 \times 10^6) \div (5 \times 10^2)$

c $(2 \times 10^5)^2$ **d** $(4 \times 10^3)^2$

e $(5 \times 10^4)^3$ **f** $(3 \times 10^5)^4$

6 Work out the following. Give your answers in standard form.

a $(2 \times 10^4) + (3 \times 10^4)$

b $(6.4 \times 10^4) + (4.6 \times 10^4)$

c $(5 \times 10^5) - (4.7 \times 10^5)$

d $(4 \times 10^6) + (3 \times 10^5)$

e $(2 \times 10^4) \times 2$ **f** $(4 \times 10^8) \div 8$

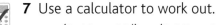 **7** Use a calculator to work out.

a $(2.42 \times 10^4) \times (2.25 \times 10^6)$

b $(3.2 \times 10^5) \times (4.56 \times 10^7)$

c $(1.4 \times 10^5)^2$

d $(4.8 \times 10^{13}) \div (1.25 \times 10^6)$

e $(7.6 \times 10^9) \div (1.6 \times 10^6)$

8 $x = 2.8 \times 10^6$, $y = 9.2 \times 10^5$

Work out each of the following. Give your answers in standard form correct to 3 significant figures.

a xy **b** y^2

c $x \div y$ **d** $\dfrac{xy}{x + y}$

9 $p = 4.8 \times 10^6$, $q = 3.6 \times 10^6$
Work out each of the following. Give your answers in standard form correct to 3 significant figures, if appropriate.

a pq **b** $(p - q)^2$

c $\dfrac{p + q}{p - q}$ **d** pq^2

Exercise 26C

1 Write the following numbers in standard form.

a 0.003   **b** 0.000 45

c 0.002 63   **d** 0.040

e 0.000 402   **f** 0.000 045

2 Write the following as ordinary numbers.

a 7×10^{-2} **b** 6.2×10^{-3}

c 4.05×10^{-4} **d** 2.4×10^{-1}

e 5.0×10^{-3}

3 Write the following in standard form.

a 60×10^{-4} **b** 540×10^{-3}

c 0.8×10^{-4} **d** 0.06×10^{-2}

e 0.005×10^0

4 Work out the following. Give your answers in standard form.

a $(4 \times 10^{-3}) \times (2 \times 10^{-4})$

b $(2.5 \times 10^2) \times (4 \times 10^{-5})$

c $(3 \times 10^{-4}) \times (4 \times 10^{-6})$

d $(4 \times 10^{-6}) \div (2 \times 10^{-4})$

e $(2 \times 10^{-5}) \div (8 \times 10^{-2})$

f $(1.5 \times 10^{-3}) \div (5 \times 10^{-7})$

5 Work out the following. Give your answers in standard form.

a $(2 \times 10^{-4})^3$ **b** $(3 \times 10^{-4})^2$

c $(4 \times 10^{-5})^2$ **d** $2 \times (2 \times 10^{-5})^2$

e $4 \times (5 \times 10^{-3})^3$

6 Work out the following. Give your answers in standard form.

a $2 \times 10^{-3} + 2 \times 10^{-3}$

b $4 \times 10^{-3} + 8 \times 10^{-4}$

c $3 \times 10^{-3} - 8 \times 10^{-4}$

d $4 \times 10^{-7} - 3 \times 10^{-8}$

 7 $x = 1.2 \times 10^{-3}$, $y = 8.8 \times 10^{-4}$
Work out the value of

a xy **b** $2y^2$

c $\dfrac{x}{y}$ **d** $\dfrac{xy}{x + y}$

8 $p = 1.2 \times 10^{-4}$, $q = 9.5 \times 10^{-5}$
Work out the value of each of the following. Give your answers correct to 3 significant figures.

a $(p + q)^2$ **b** pq^2

c $\dfrac{p - q}{p + q}$ **d** $\sqrt{pq}$

9 The mass of a hydrogen atom is 1.67×10^{-24} grams. Work out the number of hydrogen atoms in 3.2 kg of hydrogen. Give your answer correct to 3 significant figures.

10 The mass of the Earth is 5.97×10^{24} kg. The mass of the Moon is 7.35×10^{22} kg. Work out the ratio of the mass of the Earth to the mass of the Moon.
Give your answer in the form $k : 1$ where k is written correct to the nearest whole number.

Exercise 26D

1 Work out

a $36^{\frac{1}{2}}$ **b** $8^{\frac{1}{3}}$ **c** $81^{\frac{1}{2}}$

d $16^{\frac{1}{4}}$ **e** $1\,000\,000^{\frac{1}{6}}$

2 Work out

a $\left(\frac{1}{8}\right)^{\frac{1}{3}}$ **b** $49^{-\frac{1}{2}}$ **c** $8^{-\frac{1}{3}}$

d $\left(\frac{1}{9}\right)^{-\frac{1}{2}}$ **e** $\left(\frac{9}{16}\right)^{\frac{1}{2}}$ **f** $\left(\frac{27}{64}\right)^{-\frac{1}{3}}$

3 Write as a single power

a $\sqrt{3}$ **b** $\sqrt[3]{5}$ **c** $(\sqrt{6})^3$

d $\dfrac{1}{\sqrt{7}}$ **e** $\dfrac{1}{\sqrt[3]{2}}$ **f** $\dfrac{1}{\sqrt[3]{2^2}}$

4 Work out

a $81^{\frac{3}{4}}$ **b** $9^{\frac{3}{2}}$ **c** $4^{\frac{5}{2}}$

d $125^{\frac{2}{3}}$ **e** $16^{-\frac{3}{4}}$ **f** $64^{\frac{2}{3}}$

Exercise 26E

1 Find the value of k.

  a $\sqrt{50} = k\sqrt{2}$ **b** $108 = k\sqrt{3}$

  c $\sqrt{12} = k\sqrt{3}$ **d** $\sqrt{75} = k\sqrt{3}$

  e $\sqrt{44} = k\sqrt{11}$

2 Expand and simplify

  a $\sqrt{5}(\sqrt{5} - 1)$ **b** $(\sqrt{5} + 1)(\sqrt{5} - 1)$

  c $(\sqrt{3} + 2)(\sqrt{3} + 1)$

  d $(\sqrt{2} + 3)(\sqrt{2} - 1)$

  e $(\sqrt{7} - 2)^2$

3 Rationalise the denominator and simplify where appropriate.

  a $\dfrac{1}{\sqrt{6}}$ **b** $\dfrac{2}{\sqrt{3}}$ **c** $\dfrac{2}{\sqrt{10}}$

  d $\dfrac{3}{\sqrt{6}}$ **e** $\dfrac{4}{\sqrt{5}}$ **f** $\dfrac{9}{\sqrt{3}}$

  g $\dfrac{6}{\sqrt{2}}$

4 Rationalise the denominator and write your answer in its simplest form.

  a $\dfrac{\sqrt{2} + 1}{\sqrt{2}}$ **b** $\dfrac{3 - \sqrt{2}}{\sqrt{2}}$ **c** $\dfrac{3 - \sqrt{3}}{\sqrt{3}}$

  d $\dfrac{4 + \sqrt{5}}{\sqrt{5}}$ **e** $\dfrac{3 - \sqrt{6}}{\sqrt{6}}$

5 The width of a rectangle is $\sqrt{8}$ cm. The length of the rectangle is $\sqrt{12}$.

  a Work out the length of a diagonal. Give your answer in the form $k\sqrt{5}$

  b Work out the area of the rectangle. Give your answer in the form $n\sqrt{6}$

6 $\sqrt{8} - \sqrt{2} = m\sqrt{2}$ Find the value of m.

7 The hypotenuse of a right angled triangle is $\sqrt{90}$ cm.

  The height of the triangle is 6 cm.
  Work out the base of the triangle. Give your answer in the form $k\sqrt{6}$ cm.

Chapter 27 Constructions, loci and congruence

Exercise 27A

1 Draw a line AB 3.5 cm long. Construct an equilateral triangle with base AB.

2 Draw a circle, radius 2.5 cm, and construct a regular hexagon inside it.

In Questions **3–9**, copy the diagrams onto centimetre squared paper.

3

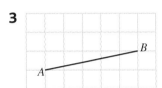

Construct the perpendicular bisector of the line AB.

4

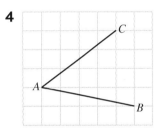

Construct the bisector of the acute angle BAC.

5

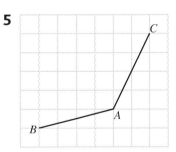

Construct the bisector of the obtuse angle BAC.

6

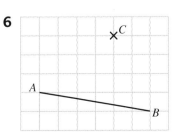

Construct the perpendicular from the point C to the line AB.

7

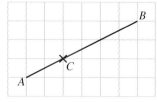

Construct the perpendicular from the point C to the line AB.

8

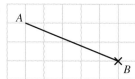

Construct the perpendicular at B to the line AB.

9

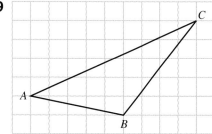

a Construct the perpendicular bisector of each of the sides of the triangle ABC.

b On a separate copy of the diagram, construct the bisector of each of the angles of the triangle ABC.

10 Construct an angle of 45°.

Exercise 27B

Copy the diagrams onto centimetre squared paper.

1 a Draw the locus of all points which are equidistant from P and Q.

b On the same diagram, draw the locus of all points which are 1 cm from P.

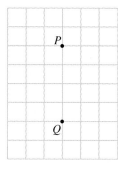

2

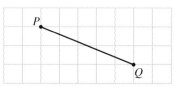

Draw the locus of all points which are 2 cm away from the line PQ.

3

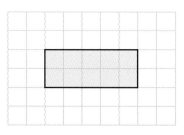

Draw the locus of the points outside the rectangle which are 2 cm from the outside of the rectangle.

4

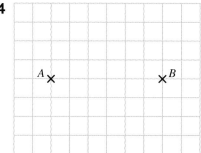

A point P is equidistant from A and B.
P is also 4 cm from A.
Find the two possible positions of P.
Mark each position with a cross.

5

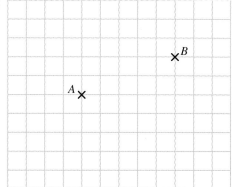

A point Q is 4 cm from A and 3 cm from B.
Find the two possible positions of Q.
Mark each position with a cross.

6

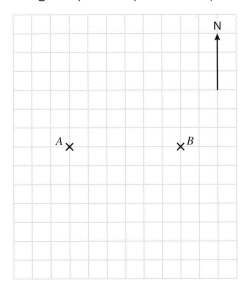

a On the grid, construct accurately, the locus of all points which are equidistant from A and B.

b On the grid, construct accurately the locus of all points which are 3 cm from C.

(1384 November 1997)

7 AB and AC are two fixed lines and D is a fixed point.
A point P is the same distance from AB and AC.
P is also 2 cm from D.
Find the *two* possible positions of P.
Mark each position with a cross.

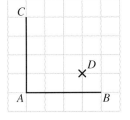

8 The grid represents part of a map.

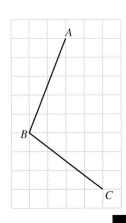

a On the grid, draw a line on a bearing of 037° from the point marked A.
The point C is on a bearing of 300° from the point marked B.
C is also 3 cm from B.

b Mark the position of the point C and label it with a letter C.

(1385 May 2002)

Exercise 27C

Copy the diagrams onto centimetre squared paper.

1

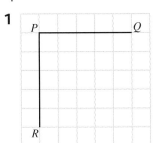

A point moves so that it is always nearer PR than PQ.
Show, by shading, the region which satisfies this condition.

2

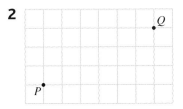

A point moves so that it is always nearer P than Q.
Show, by shading, the region which satisfies this condition.

3 A point moves so that it is always nearer AB than AC.
Show, by shading, the region which satisfies this condition.

4

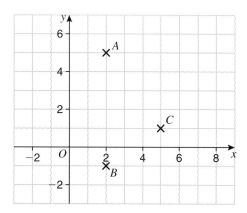

A point moves so that it is always less than 1 cm from the line AB and also less than 2 cm from the point C.
Show, by shading, the region which satisfies both these conditions.

5 The diagram shows three points A, B and C on a centimetre grid.

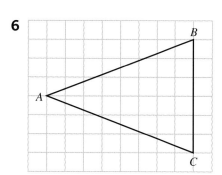

a Draw the locus of points whish are equidistant from A and B.

b Draw the locus of points which are 3 cm from C.

c On the grid, shade the region in which points are nearer to A than B and also less than 3 cm from C.

(1388 January 2004)

6

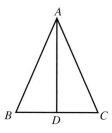

The diagram represents a triangular garden ABC.

The scale of the garden is 1 cm represents 1 m. A tree is to be planted in the garden so that it is nearer to AB than to AC, within 5 m of the point A.
Shade the region where the tree may be planted.

Exercise 27D

1 Use the given measurements to make an accurate drawing of triangle ABC with AB as base.

a $AB = 5.2$ cm, $AC = 5.8$ cm, angle $A = 54°$.

b $AB = 5.9$ cm, $BC = 6.2$ cm, $AC = 4.7$ cm.

c $AB = 6.1$ cm, $BC = 5.5$ cm, angle $B = 107°$.

d $AB = 5.4$ cm, angle $A = 55°$, angle $B = 69°$.

e $AB = 5.7$ cm, angle $A = 90°$, $BC = 6.2$ cm.

2 Make accurate drawings of two triangles with the given measurements.

a Angle $A = 62°$, angle $B = 52°$, angle $C = 66°$.

b $AB = 6.8$ cm, angle $A = 56°$, $BC = 5.9$ cm.

c $AB = 6.5$ cm, $AC = 4.8$ cm, angle $B = 43°$.

Exercise 27E

1 In the isosceles triangle ABC, $AB = AC$ and AD is perpendicular to BC.

Prove that triangles ABD and ACD are congruent.

2 Two straight lines, AB and CD, cross at E.
$AE = CE$ and angle DAE = angle BCE.

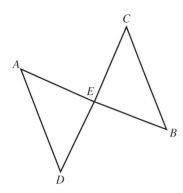

Prove that triangles AED and CEB are congruent.

3

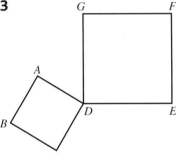

Diagram **NOT** accuratey drawn

$ABCD$ and $DEFG$ are squares.
Prove that triangle CDG and triangle ADE are congruent.
(1387 November 2003)

4

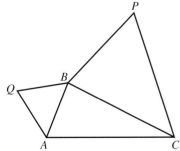

Diagram **NOT** accuratey drawn

In the diagram, triangle ABQ and triangle BCP are equilateral.
Explain why, triangle ABP and triangle QBC are congruent.
(1385 November 1998)

Chapter 28 Further factorising, simplifying, completing the square and algebraic proof

Exercise 28A

1 a Write down two numbers whose product is $+12$ and whose sum is $+7$

  b Hence factorise $2x^2 + 7x + 6$

2 a Write down two numbers whose product is -24 and whose sum is $+10$

  b Hence factorise $8x^2 + 10x - 3$

3 a Write down two numbers whose product is -48 and whose sum is -2

  b Hence factorise $16y^2 - 2y - 3$

4 Factorise $2y^2 - 11y + 12$
(1386 November 1999)

5 Factorise these expressions.

  a $10x^2 + 7x + 1$ **b** $15x^2 - 14x - 1$
  c $3x^2 - x - 2$ **d** $4x^2 + 11x + 6$
  e $2x^2 - x - 36$ **f** $36x^2 + 17x + 2$
  g $5x^2 + 8x - 4$ **h** $5x^2 - 13x - 6$
  i $3y^2 + 7y - 6$ **j** $12y^2 - 13y - 4$
  k $2x^2 + 17xy + 8y^2$ **l** $3y^2 - 5xy - 8x^2$

6 Factorise $2x^2 + 19x - 33$
(1386 November 2000)

7 Factorise completely.

  a $18x^2 + 27x - 18$
  b $12 + 58x + 18x^2$
  c $12y^3 - 15y^2 - 18y$

8 a Factorise $2y^2 + 13y + 21$

  b Hence, or otherwise, factorise $2(3x - 1)^2 + 13(3x - 1) + 21$

Exercise 28B

1 Simplify

a $\dfrac{12p^5}{3p^3}$

b $\dfrac{3(2y+3)}{(2y+3)^3}$

c $\dfrac{(x+2)(x+3)}{x(2+x)}$

d $\dfrac{2-3x}{3x-2}$

e $\dfrac{(4-x)(1-x)}{(x-4)(x+1)}$

2 Simplify fully $\dfrac{3(2x+1)}{4x^2-1}$

(1388 March 2006)

3 Simplify fully $\dfrac{x^2+5x+6}{x^2+2x}$

(1387 June 2006)

4 Simplify these expressions fully

a $\dfrac{4x+6}{2x}$

b $\dfrac{8y^2+6y}{2y}$

c $\dfrac{5x+20}{6x+24}$

d $\dfrac{4-6p}{6p^2-4p}$

e $\dfrac{x^2-4x}{x^2-16}$

f $\dfrac{25-q^2}{10q+2q^2}$

g $\dfrac{2x^2-8}{x^2-3x+2}$

h $\dfrac{2x^2+8x+6}{2x^2+7x+3}$

i $\dfrac{1-x^2}{x^2-4x+3}$

j $\dfrac{4x^2-25}{4x^2-12x+5}$

k $\dfrac{2x-9x^2}{18x^2+5x-2}$

l $\dfrac{18x^2-50}{9x^2-3x-20}$

5 Simplify fully

a $\dfrac{2y^2-8}{4y^2+12y+8} \times \dfrac{y+1}{y-2}$

b $\dfrac{3-4x+x^2}{2x^2+x-3} \times \dfrac{9-4x^2}{2x^2-7x+3}$

Exercise 28C

1 Write $\dfrac{3}{4x} + \dfrac{5}{6x} - \dfrac{1}{3x}$ as a single fraction in its simplest form.

2 Write $\dfrac{4}{7x+14} - \dfrac{1}{2x+4}$ as a single algebraic fraction.

3 Express $\dfrac{1}{x+4} - \dfrac{1}{x+6}$ as a single fraction.

4 Write $\dfrac{3}{5} - \dfrac{2}{3x+4}$ as a single fraction.

5 a Factorise x^2-6x

b Simplify $\dfrac{6}{x^2-6x} + \dfrac{1}{x}$

6 Simplify $\dfrac{2}{x-2} - \dfrac{8}{x^2-4}$

7 Write $\dfrac{1}{6x-5} - \dfrac{1}{6x-1}$ as a single fraction.

8 Simplify $\dfrac{3}{x+2} + \dfrac{9}{x^2+x-2}$

9 Show that

$$\dfrac{x+3}{x+1} - \dfrac{x+1}{x+2} + \dfrac{1}{x^2+3x+2} = \dfrac{3}{x+1}$$

10 Write $\dfrac{3}{8x^2-2x-1} + \dfrac{1}{8x^2+6x+1}$ as a single fraction in its simplest form.

Exercise 28D

1 Write in the form $(x+p)^2 + q$

a x^2+8x

b x^2+14x

c x^2+16x

d x^2-20x

e x^2-2x

f x^2-10x

g $x^2 + 4x + 23$ **h** $x^2 + 18x + 73$
i $x^2 + 22x - 9$ **j** $x^2 - 14x + 50$
k $x^2 - 6x - 2$ **l** $x^2 - 12x + 26$

2 Write in the form $a(x + p)^2 + q$
  a $2x^2 + 8x$ **b** $3x^2 + 18x + 5$
  c $8x^2 - 160x + 560$

3 Write in the form $p - (x + q)^2$.
  a $1 + 6x - x^2$ **b** $29 - 10x - x^2$
  c $1 - x - x^2$

4 For all values of x,
  $x^2 + 14x + 37 = (x + p)^2 + q$.
  a Find the value of p and the value of q.
  b Write down the minimum value of $x^2 + 14x + 37$.

5 The curve with equation $y = x^2 - 12x + 32$ has a minimum point.
  Find the coordinates of the minimum point.

6 The curve with equation $y = 15 + 4x - x^2$ has a maximum point.
  a Write the expression $15 + 4x - x^2$ in the form $p - (x + q)^2$.
  b Hence find the coordinates of the maximum point.

7 The curve with equation $y = 3x - 3x^2$ has a maximum point.
  Find the coordinates of the maximum point.

Exercise 28E

1 Prove algebraically that the sum of any five consecutive integers is always a multiple of 5.

2 Prove algebraically that the sum of any three consecutive odd numbers is always a multiple of 3.

3 Prove algebraically that the sum of the squares of any three consecutive integers is 1 less than a multiple of 3.

4 a Factorise $p^2 - q^2$.
  Here is a sequence of numbers
  0, 3, 8, 15, 24, 35, 48, ...
  b Write down an expression for the nth term of this sequence.
  c Show algebraically that the product of **any** two consecutive terms of the sequence
  0, 3, 8, 15, 24, 35, 48, ...
  can be written as the product of four consecutive integers.

Chapter 29 Circle geometry

Exercise 29A

In Questions **1–4**, each diagram shows a circle, centre O.
Calculate the size of each of the angles marked with a letter.
The diagrams are **NOT** accurately drawn.

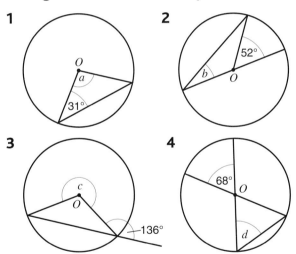

In Questions **5** and **6**, give reasons for your answers.

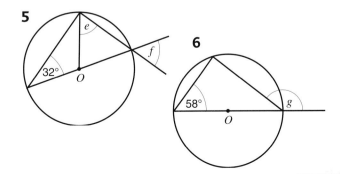

Exercise 29B

The diagrams are **NOT** accurately drawn.

1 PT is a tangent at T to a circle, centre O.
Angle $OPT = 28°$.

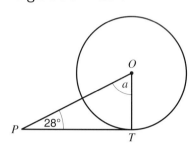

Find the size of angle a.
Give reasons for your answer.

2 PA is a tangent at A
to a circle, centre O.
B is a point on the
circumference of the
circle.
POB is a straight line.
Find the size of angle y.

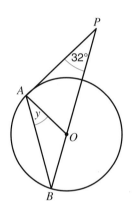

3 PA is a tangent at A to a circle, centre O.
AB is a chord of the circle.
Calculate the size of each of the angles
marked with letters.

a

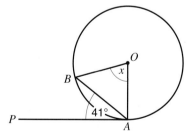

b

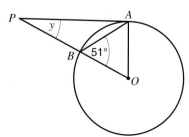

4 AB is a chord of a circle,
centre O.
M is the midpoint of
AB.
Angle $AOM = 29°$.
Find the size of angle
OAB.
Give reasons for your answer.

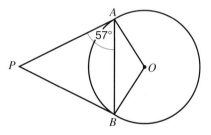

5 PA and PB are tangents to a circle,
centre O.

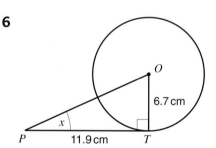

Find the size of

a angle APB

b angle AOB.

6

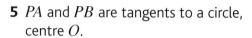

In the diagram, T is a point on the circle,
centre O.
PT is a tangent to the circle at T.

a Angle OTP is a right angle.
Give a reason why.

The radius of the circle is 6.7 cm.
$PT = 11.9$ cm.

b Calculate the length of OP.
Give your answer correct to 3 significant
figures.

c Calculate the size of angle x.
Give your answer correct to 1 decimal
place.

Exercise 29C

1

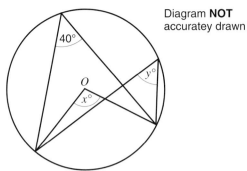

Diagram **NOT** accuratey drawn

In the diagram O is the centre of the circle.
Write down the value of

a x **b** y.   *(1384 November 1994)*

2 A, B, C and D are points on a circle.
Angle $BAC = 40°$. Angle $DBC = 55°$.

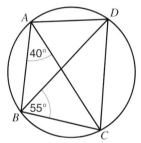

Diagram **NOT** accuratey drawn

a   **i**   Find the size of angle DAC.
    ii   Give a reason for your answer.
b   **i**   Calculate the size of angle DCB.
    ii   Give reasons for your answer.
c   Is BD a diameter of the circle?
    Give a reason for your answer.
             (4400 November 2004)

3

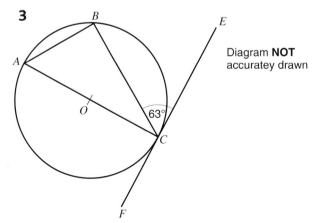

Diagram **NOT** accuratey drawn

In the diagram, A, B and C are points on a circle, centre O.
Angle $BCE = 63°$.
FE is a tangent to the circle at point C.

a   Calculate the size of angle ACB.
    Give reasons for your answer.
b   Calculate the size of angle BAC.
    Give reasons for your answer.
             (1387 June 2003)

4

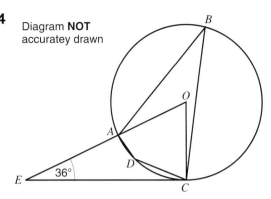

Diagram **NOT** accuratey drawn

A, B, C and D are points on the circumference of a circle, centre O.
A tangent is drawn from E to touch the circle at C.
Angle $AEC = 36°$. EAO is a straight line.

a   Calculate the size of angle ABC.
    Give reasons for your answer.
b   Calculate the size of angle ADC.
    Give reasons for your answer.
             (1385 June 2001)

5 The diagram shows a circle centre O.
PQ and PR are tangents to the circle at P and Q respectively.
S is a point on the circle.
Angle $PSR = 70°$.
$PS = SR$.

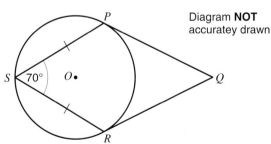

Diagram **NOT** accuratey drawn

a   **i**   Calculate the size of angle PQR.
    ii   Give a reason for your answer.
b   **i**   Calculate the size of angle SPQ.
    ii   Explain why $PQRS$ cannot be a cyclic quadrilateral.
             (1384 November 1997)

6

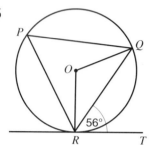

Diagram **NOT** accuratey drawn

P, Q and R are points on a circle.
O is the centre of the circle.
RT is the tangent to the circle at R.
Angle $QRT = 56°$.

a   Find
   i   the size of angle RPQ
   ii   the size of angle ROQ.

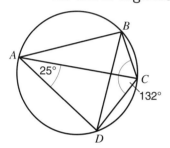

Diagram **NOT** accuratey drawn

A, B, C and D are points on a circle.
AC is a diameter of the circle.
Angle $CAD = 25°$ and angle $BCD = 132°$.

b   Calculate
   i   the size of angle BAC
   ii   the size of angle ABD.

(1386 May 2002)

Chapter 30 Quadratic equations

Exercise 30A

1 Solve these equations.
   a $y^2 = 36$ **b** $2y^2 = 18$
   c $2y^2 = 200$ **d** $4y^2 = 36$

2 Solve these equations.
   a $(x + 4)(x - 6) = 0$
   b $(x - 4)(x - 8) = 0$
   c $(2x + 3)(x - 4) = 0$
   d $(x - 2)(2x - 6) = 0$
   e $x(x - 6) = 0$

3 Solve these equations.
   a $x^2 - 4x = 0$ **b** $x^2 + 3x = 0$
   c $2x^2 - 5x = 0$ **d** $3x^2 - 4x = 0$
   e $4x^2 - 4x = 0$

4 Solve these equations.
   a $x^2 + 5x + 4 = 0$
   b $x^2 + 6x + 5 = 0$
   c $x^2 + 14x + 24 = 0$
   d $x^2 - 6x + 5 = 0$
   e $x^2 - 9x + 8 = 0$
   f $x^2 - 8x + 15 = 0$
   g $x^2 - 4x - 5 = 0$
   h $x^2 - 5x - 6 = 0$
   i $x^2 - x - 6 = 0$

5 Solve these equations
   a $2x^2 - 3x - 2 = 0$
   b $3x^2 - 4x + 1 = 0$
   c $2x^2 - 3x + 1 = 0$
   d $4x^2 - 11x - 3 = 0$
   e $3x^2 - 8x - 3 = 0$
   f $5x^2 - 3x - 2 = 0$

6 Solve these equations
   a $x^2 - 3x = 4$ **b** $x^2 - 2x = 3$
   c $x^2 + 6x = 7$ **d** $x^2 - 2x = 8$
   e $x = 12 - x^2$ **f** $8x + x^2 = 20$

7 Solve these equations
   a $2x^2 - 3x = 5$ **b** $2x^2 + 5x = 7$
   c $3x^2 - 4x = 4$ **d** $5x^2 - 3x = 8$
   e $2x^2 = 5x - 2$ **f** $2x = 5x^2 - 3$

8 Solve these equations
   a $(x + 3)(x + 5) = 24$
   b $(x + 3)(x - 4) = 8$
   c $(x - 2)(x + 5) = 30$
   d $(2x + 1)(x + 1) = 1$
   e $(2x + 3)(x - 1) = 7$
   f $(2x + 3)(2x + 1) = -1$

Exercise 30B

1 Solve the following equations by completing the square.
Give your solutions in the form $a \pm b\sqrt{c}$ where a, b and c are integers.

a $x^2 + 6x - 12 = 0$ **b** $x^2 + 10x - 1 = 0$

c $x^2 - 8x - 13 = 0$ **d** $x^2 - 8x - 17 = 0$

e $x^2 - 4x + 2 = 0$

2 Solve the following equations. Give your answers in the form $a \pm b\sqrt{c}$ where a, b and c are integers.

a $(x - 3)^2 = 5$ **b** $(x - 2)^2 = 6$

c $(x + 1)^2 = 3$ **d** $(x - 3)^2 - 7 = 0$

e $(x + 4)^2 - 8 = 0$

Exercise 30C

1 Solve these equations. Give your answers correct to 3 significant figures.

a $x^2 + 5x + 3 = 0$ **b** $x^2 - 6x - 2 = 0$

c $x^2 - 7x + 4 = 0$ **d** $x^2 - 5x - 3 = 0$

e $x^2 - 4x - 6 = 0$ **f** $x^2 + 6x - 3 = 0$

2 Solve these equations. Give your answers correct to 3 significant figures.

a $2x^2 - 3x - 4 = 0$

b $3x^2 - 2x - 2 = 0$

c $2x^2 - 5x - 4 = 0$

d $2x^2 - 7x + 2 = 0$

e $4x^2 - 6x + 1 = 0$

f $2x^2 - x - 4 = 0$

g $2x^2 - 6x + 3 = 0$

h $5x^2 - 6x - 3 = 0$

i $6x - x^2 - 2 = 0$

3 Solve these equations. Give your answers correct to 3 significant figures.

a $x(x + 2) - 4 = 0$ **b** $x(x - 3) - 5 = 0$

c $x^2 - 6x = 10$ **d** $x^2 + 4x = 30$

e $2x^2 + 6x = 15$ **f** $x^2 - 15 = 4x$

g $8x - x^2 = 2$ **h** $x + 3 = x^2$

i $13x - 5x^2 = 4$

Exercise 30D

1 Solve these equations

a $\dfrac{4}{x} + \dfrac{4}{x + 2} = 3$ **b** $\dfrac{3}{x} + \dfrac{4}{x + 1} = 2$

c $\dfrac{2}{x - 1} + \dfrac{1}{x + 2} = 0$

d $\dfrac{6}{x + 1} - \dfrac{4}{x + 2} = 1$

2 Solve these equations. Give your solutions correct to 3 significant figures.

a $\dfrac{1}{x} + \dfrac{3}{x + 2} = 3$

b $\dfrac{2}{x - 1} + \dfrac{1}{x + 1} = 3$

c $\dfrac{1}{2x} + \dfrac{1}{x + 2} = 2$

d $\dfrac{3}{x} - \dfrac{3}{x - 2} = 1$

3 Solve these equations

a $\dfrac{3}{2x - 1} - x = 2$ **b** $\dfrac{4}{x} - \dfrac{2x}{3} = \dfrac{2}{3}$

c $\dfrac{x + 3}{2} - \dfrac{2}{x} = 3$ **d** $\dfrac{x}{x + 1} - \dfrac{1}{2x} = \dfrac{5}{12}$

Exercise 30E

1 A rectangle has a width of x cm and a length of $(x + 4)$ cm. The area of the rectangle is 77 cm².

a Show that $x^2 + 4x = 77$

b Solve the equation $x^2 + 4x = 77$

c Find the width and the length of the rectangle.

2 The product of two consecutive numbers is 72. Let x be the smaller of the two consecutive numbers.

a Show that $x^2 + x = 72$

b Solve the equation and write down the values of the two consecutive numbers.

3 The lengths of the two shortest sides of a right-angled triangle are x cm and $(x + 3)$ cm respectively. The length of the hypotenuse is 15 cm.

   a Show that $x^2 + 3x = 108$

   b Solve the equation $x^2 + 3x = 108$

   c Write down the lengths of the two shortest sides of the triangle.

4 The height of a triangle is 4 cm more than the base. The area of the triangle is 16 cm². Let x cm be the base of the triangle.

   a Show that $x(x + 4) = 32$

   b Solve the equation $x(x + 4) = 32$

   c Find the height of the triangle.

5 The sum of the squares of 3 consecutive numbers is 50.
Let x be the smallest of the numbers.

   a Show that $3x^2 + 6x + 5 = 50$

   b Solve the equation.

6 The diagram shows a hexagon. All the angles are right angles. The area of the hexagon is 59 cm².

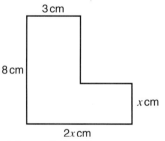

   a Show that $2x^2 - 3x - 35 = 0$

   b Solve the equation

   c Write down the length of the longest side of the hexagon.

7 The diagram shows a trapezium. All the measurements are in cm.
The area of the trapezium is 112 cm².

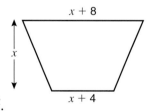

   a Show that $x^2 + 6x = 112$

   b Solve the equation

   c Write down the length of the longest side of the trapezium.

8 The average speed of a car on a journey of 240 km is x kilometres per hour.
The average speed of the car on the return journey is 20 kilometres per hour faster.
The total time for the journeys there and back is 10 hours.

   a Explain why $\dfrac{240}{x} + \dfrac{240}{x + 20} = 10$

   b Show that $x^2 - 28x - 480 = 0$

   c Solve the equation and find the average speed of the car on the return journey.

Chapter 31 Pythagoras' theorem and trigonometry (2)

Exercise 31A

Where necessary, give lengths correct to 3 significant figures and angles correct to one decimal place.

1 The diagram shows a cuboid. A, B, C, D, E and F are six vertices of the cuboid.
$AB = 12$ cm, $BC = 15$ cm, $CE = 5$ cm

Diagram **NOT** accuratey drawn

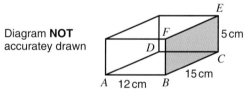

   a Calculate the length of
     i AF,   **ii** BE,   **iii** AE.

   b Calculate the size of
     i angle FAB, **ii** angle BEF, **iii** angle EAC.

2 $ABCDEF$ is a triangular prism.
The rectangular plane $CDEF$ is horizontal and the rectangular plane $ABFE$ is vertical.
$AE = 4$ cm, $ED = 8$ cm, $DC = 10$ cm.

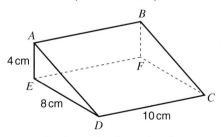

   a Calculate the length of
     i FD,   **ii** AF,   **iii** AC.

b Calculate the size of
i angle *ADE*, ii angle *BFA*, iii angle *ACE*.

3 The diagram shows a pyramid.
The base, *ABCD*, is a horizontal square of
side 10 cm. The vertex, *V*, is vertically above
the midpoint, *M*, of the base.
VM = 12 cm.

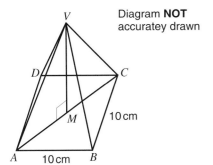

Diagram **NOT**
accuratey drawn

Calculate the size of angle *VAM*.
(4400 November 2005)

Exercise 31B

1 The diagram represents a cuboid
ABCDEFGH.
AB = 5 cm, *BC* = 7 cm, *AE* = 3 cm.

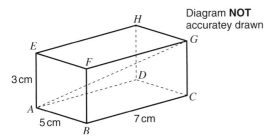

Diagram **NOT**
accuratey drawn

a Calculate the length of *AG*.
Give your answer correct to 3 significant
figures.

b Calculate the size of the angle between
AG and the face *ABCD*.
Give your answer correct to 1 decimal
place.
(1387 November 2004)

2 The diagram represents a prism.
AEFD is a rectangle. *ABCD* is a square.
EB and *FC* are perpendicular to plane
ABCD.
AB = 60 cm. *AD* = 60 cm.
Angle *ABE* = 90°. Angle *BAE* = 30°

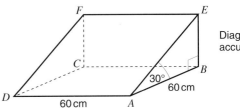

Diagram **NOT**
accuratey drawn

Calculate the size of the angle that the line
DE makes with the plane *ABCD*.
Give your answer correct to 1 decimal place.
(1387 May 2004)

3 The diagram shows a square-based pyramid.
The lengths of the sides of the square base,
ABCD, are 12 cm and the base is on a
horizontal plane. The centre of the base is *M*
and the vertex of the pyramid is *V* so that
VM is vertical and *VM* = 18 cm.
The midpoint of *BC* is *N*.

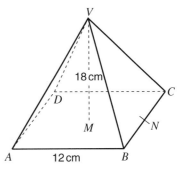

a Calculate the length of
i *AM*, ii *VA*, iii *VN*.
Give each answer correct to 3 significant
figures.

b Calculate the size of the angle
i between *VB* and the plane *ABCD*,
ii between *VN* and the plane *ABCD*.
Give each answer correct to 1 decimal
place.

Exercise 31C

1 Use a calculator to find the value, correct to
3 decimal places if necessary, of
a sin 150° **b** cos 110°
c tan 137° **d** cos 317°
e tan 244° **f** sin 286.4°
g cos 238.8°

2 x is an obtuse angle. Find the value of x, in degrees, when

   a $\sin x = 0.6$

   b $\cos x = -0.38$

   c $\tan x = -1.47$.

3 Here is the graph of $y = \sin x°$ for $0 \leqslant x \leqslant 180$.

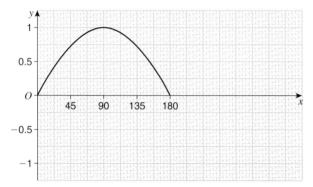

   a Accurately copy this graph and continue the curve to show the graph of $y = \sin x°$ for $0 \leqslant x \leqslant 360$.

   b Using your graph, or otherwise, find estimates of the solutions in the interval $0 \leqslant x \leqslant 360$ of the equation
   i $\sin x° = 0.7$ **ii** $\sin x° = -0.4$

4 The diagram shows part of the curve $y = \cos x°$ for $0 \leqslant x \leqslant 360$.

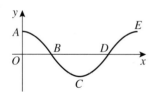

Write down the coordinates of the point
i A, **ii** B, **iii** C, **iv** D, **v** E.

5 Here is a sketch of part of the graph of $y = \sin x°$.

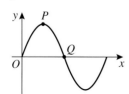

Write down the coordinates of
i P, **ii** Q.   *(1387 June 2005)*

Exercise 31D

1 Work out the area of each of these triangles. Give each answer correct to 3 significant figures.

   a

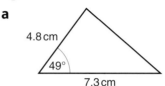

   b

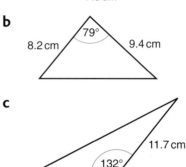

   c

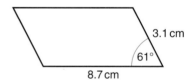

2 Calculate the area of the parallelogram. Give your answer correct to 3 significant figures.

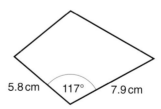

3 Here is a kite. Calculate the area of the kite. Give your answer correct to 1 decimal place.

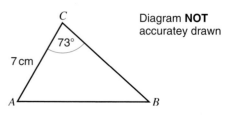

4 The area of triangle ABC is 33.5 cm².

Diagram **NOT** accuratey drawn

Calculate the length of BC.
Give your answer correct to nearest cm.

5 OPQ is a sector of a circle, centre O and radius 12 cm. The size of angle POQ is 83°.

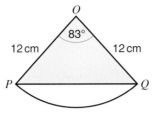

a Calculate the area of triangle OPQ.

b Calculate the area of sector OPQ.

c Hence calculate the area of the segment shown unshaded in the diagram.

Give each answer, in cm², correct to 1 decimal place.

Exercise 31E

1 Calculate the lengths of the sides marked with letters in these triangles. Give each answer correct to 3 significant figures.

a

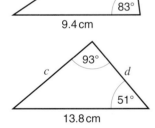

b

c

2 Calculate the size of each of the acute angles marked with a letter. Give each answer correct to 1 decimal place.

a

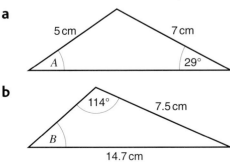

b

c

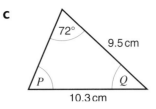

3 $BC = 9.4$ cm.
Angle $BAC = 123°$. Angle $ABC = 35°$.

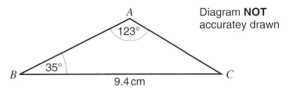

Diagram **NOT** accuratey drawn

a Calculate the length of AC. Give your answer correct to 3 significant figures

b Calculate the area of triangle ABC. Give your answer correct to 3 significant figures. *(4400 May 2005)*

4 In triangle ABC, angle $ABC = 60°$, angle $ACB = 40°$, $BC = 12$ cm.

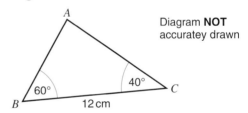

Diagram **NOT** accuratey drawn

Work out the length of AB.

5 The diagram shows a vertical tower DC on horizontal ground ABC.
ABC is a straight line.
The angle of elevation of D from A is 28°.
The angle of elevation of D from B is 54°.
$AB = 25$ m.

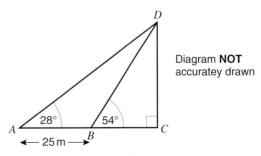

Diagram **NOT** accuratey drawn

Calculate the height of the tower.
Give your answer correct to 3 significant figures. *(1387 June 2006)*

Exercise 31F

1 Calculate the lengths of the sides marked with letters in these triangles. Give each answer correct to 3 significant figures.

a

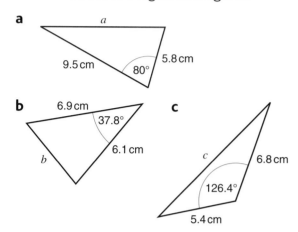

b 6.9 cm

c

2 Calculate the size of each of the angles marked with a letter. Give each answer correct to 1 decimal place.

a

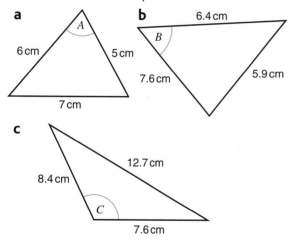

b 6.4 cm

c

3

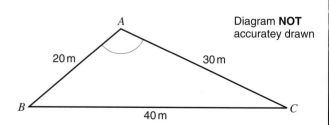

Diagram **NOT** accuratey drawn

a Calculate the length of AB.
Give your answer, in centimetres, correct to 3 significant figures.

b Calculate the size of angle ABC.
Give your answer correct to 3 significant figures.
(1385 June 2001)

4 In triangle ABC,
$AB = 9$ cm, $BC = 15$ cm, angle $ABC = 110°$.

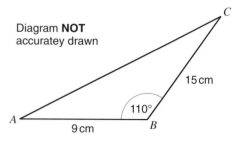

Diagram **NOT** accuratey drawn

Calculate the perimeter of the triangle.
Give your answer correct to the nearest cm.

Exercise 31G

1 a A farmer arranges 90 m of fencing in the form of an isosceles triangle, with two sides of length 35 m and one side of length 20 m.

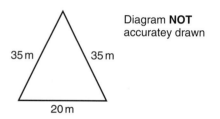

Diagram **NOT** accuratey drawn

Calculate the area enclosed by the fencing.
Give your answer correct to 3 significant figures.

b Later the farmer moves the fencing so that it forms a different triangle, ABC.
$AB = 20$ m $BC = 40$ m $CA = 30$ m

Diagram **NOT** accuratey drawn

Calculate the size of angle BAC.
Give your answer correct to 1 decimal place.

(4400 May 2006)

2 $AB = 3.2$ cm.
$BC = 8.4$ cm
The area of triangle ABC is 10 cm^2.
Calculate the perimeter of triangle ABC.
Give your answer correct to 3 significant figures.
(1387 June 2004)

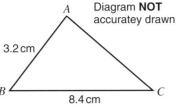

Diagram **NOT** accuratey drawn

A

3.2 cm

B 8.4 cm C

3 A straight road UW has been constructed to by-pass a village V.
The original straight roads UV and VW are 4 km and 5 km in length respectively.
V lies on a bearing of 052° from U.
W lies on a bearing of 078° from V.
The average speed on the route UVW through the village is 30 kilometres per hour.
The average speed on the by-pass route UW is 65 kilometres per hour.

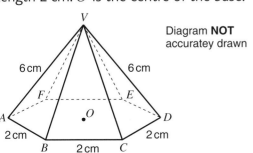

Diagram **NOT** accuratey drawn

Calculate the time saved by using the by-pass route UV.
Give your answer to the nearest minute.
(1384 June 1995)

4 The diagram shows shows a pyramid. The apex of the pyramid is V. Each of the sloping edges is of length 6 cm. The base of the pyramid is a regular hexagon with sides of length 2 cm. O is the centre of the base.

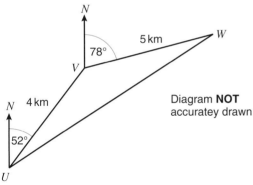

Diagram **NOT** accuratey drawn

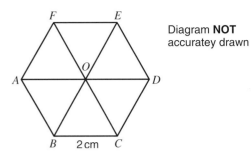

Diagram **NOT** accuratey drawn

a Calculate the height of V above the base of the pyramid.
Give your answer correct to 3 significant figures.

b Calculate the size of angle DVA.
Give your answer correct to 3 significant figures.

c Calculate the size of angle AVC.
Give your answer correct to 3 significant figures.
(1387 June 2005)

Chapter 32 Simultaneous linear and quadratic equations

Exercise 32A

1 Solve these simultaneous equations.
  a $y = 3x$ and $y = x^2$
  b $y = 4x$ and $y = 2x^2$
  c $y = 3 + x$ and $y = x^2 + 1$
  d $y = x - 5$ and $y = x^2 - 5x$
  e $y = 8 + x$ and $y = x^2 - x$
  f $y = x + 2$ and $y = 4 - x^2$

2 Solve.
  a $y - x = 4$ and $y = x^2 + 2$
  b $y - 2x = 3$ and $y = x^2 + 3$
  c $y - 3x = 1$ and $y = x^2 + 3$
  d $y - x = 4$ and $y = 2x^2 - 6x + 7$
  e $y + x = 4$ and $y = x^2 + 4$
  f $y + 3x = 4$ and $y = 2x^2 - 1$
  g $2x + y = 1$ and $y = x^2 - 2$
  h $x + 2y = 1$ and $y = x^2 + 2x - 1$
  i $4x - 3y = 10$ and $y = 2x^2 - 4$
  j $2x - 3y = 4$ and $y = 2x^2 - x - 1$

3 Find the coordinates of the points of intersection of these lines and quadratic curves.

a $y = 4$ and $y = x^2 + 3x$

b $y = 8$ and $y = x^2 - 2x$

c $y = -4$ and $y = x^2 + 4x$

d $y = -4$ and $y = x^2 - 5x$

4 Find the coordinates of the points of intersection of these lines and quadratic curves.

a $y = x + 3$ and $y = 2x^2$

b $y = 3x + 3$ and $y = x^2 - 1$

c $y = 5x + 3$ and $y = 2x^2$

d $y = 2x + 1$ and $y = 3x^2$

e $y = x + 2$ and $y = 3x^2$

f $y - x = 3$ and $y = 4x^2$

Exercise 32B

1 In each case sketch the locus of the point which moves so that it is equidistant from the given points. Give the equation of each locus.

a $(4, 0)$ and $(8, 0)$ b $(0, 4)$ and $(0, -2)$

c $(2, 0)$ and $(0, 2)$ d $(-3, 0)$ and $(0, 3)$

2 Write down the equations of these circles.

a Centre $(0, 0)$ radius 5

b Centre $(0, 0)$, radius 10

3 In each case sketch the locus of the point which moves so that it is equidistant from the given lines. Give the equation of each locus.

a $y = 2$ and $y = 6$ b $y = 4$ and $y = -2$

c $x = 4$ and $x = 7$ d $x = -2$ and $x = 6$

4 Write down the equation of the locus of the point P which moves according to both the following rules

i the coordinates of P are both positive and

ii P moves so that it is equidistant from the x-axis and the y-axis.

5 P is the point (x, y).

a Write down an expression for the distance of P from the line $x = -1$

b Write down an expression for the distance of P from the point $(0, 0)$.

P moves so that its distance from the line $x = -1$ is the same as its distance from the point $(0, 0)$.

c Find the equation of the locus of P. Give your answer in its simplest form.

6 a Draw on graph paper the circle with equation $x^2 + y^2 = 25$

b Using the same axes, draw the straight line with equation $y = x + 2$

c Hence find estimates of the solutions to the simultaneous equations $x^2 + y^2 = 25$ and $y = x + 2$

7 Draw suitable graphs to find estimates of the solutions to these pairs of simultaneous equations.

a $x^2 + y^2 = 16$ and $y = x - 2$

b $x^2 + y^2 = 64$ and $2x + 3y = 12$

8 By drawing suitable graphs, solve these simultaneous equations.

a $x^2 + y^2 = 25$ and $y = 3 - x$

b $x^2 + y^2 = 81$ and $x + y = 2$

Exercise 32C

1 Solve these simultaneous equations.

a $x^2 + y^2 = 10$ b $x^2 + y^2 = 26$
  $y = x + 2$ $y = 2x + 3$

c $x^2 + y^2 = 50$ d $x^2 + y^2 = 41$
  $y = x + 6$ $y = 2x - 3$

2 Find algebraically the coordinates of the points of intersection of each circle and straight line.

a $x^2 + y^2 = 40$ b $x^2 + y^2 = 25$
  $y = x + 1$ $y = x + 2$

c $x^2 + y^2 = 17$ d $x^2 + y^2 = 11$
  $y = x - 1$ $y = 2x - 1$

3 Solve these simultaneous equations. Give your answers correct to 3 significant figures.

 a $x^2 + y^2 = 40$ **b** $x^2 + y^2 = 25$
   $y = x + 1$ $y = x + 2$

 c $x^2 + y^2 = 20$ **d** $x^2 + y^2 = 18$
   $y = 4 - x$ $y = 2x + 2$

4 $x^2 + y^2 = 16$
  $y = 4$

 i By eliminating, show that these simultaneous equations have a solution.
 ii On the same axes, sketch the graphs of $x^2 + y^2 = 16$ and $y = 4$ and interpret your solution.

5 $x^2 + y^2 = 5$
  $y = 2x - 5$

 i Solve the equations
 ii On the same axes, sketch the graphs of $x^2 + y^2 = 5$ and $y = 2x - 5$ and interpret your solution.

Chapter 33 Similar shapes

Exercise 33A

1 Triangles PQR and STU are similar.

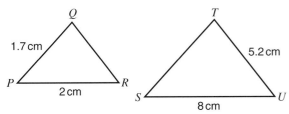

 Calculate the length of **a** ST, **b** QR.

2 Triangles ABC and DEF are similar.

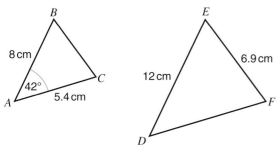

 a Find the size of angle EDF.
 Calculate the length of **b** DF, **c** BC.

3

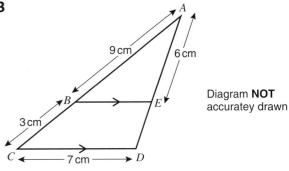

Diagram **NOT** accuratey drawn

 BE is parallel to CD. $AB = 9$ cm, $BC = 3$ cm, $CD = 7$ cm, $AE = 6$ cm. Calculate the length of ED.

        (1387 June 2005)

4

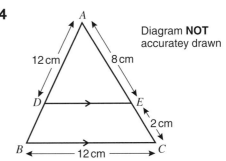

Diagram **NOT** accuratey drawn

 DE is parallel to BC.
 ADB and AEC are straight lines.
 $AD = 12$ cm. $BC = 12$ cm.
 $AE = 8$ cm. $EC = 2$ cm.
 Calculate the length of **a** DE **b** DB.
      (1385 November 1998)

In Questions **5 and 6**, AB is parallel to DE. ACE and BCD are straight lines.

5 Calculate the length of

 a BC

 b CE.

6 Calculate the length of

 a DE

 b BC.

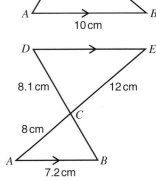

Exercise 33B

1 Rectangles **P** and **Q** are similar.

Work out the value of x.

2 Rectangles **R** and **S** are similar.

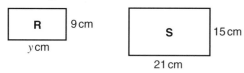

Work out the value of y.

3

Are these two rectangles mathematically similar?
You must show working to justify your answer.   *(4400 May 2006)*

4

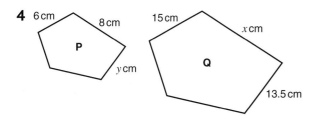

Pentagons **P** and **Q** are similar.
Calculate the value of   **a** x **b** y.

5

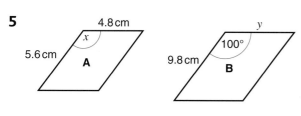

Diagram **NOT** accuratey drawn

Parallelogram **A** is mathematically similar to parallelogram **B**.

a   Find the size of the angle marked x.

b   Find the length of the side marked y.

6

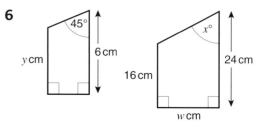

Diagram **NOT** accuratey drawn

The small trapezium and the big trapezium are mathematically similar.

a   Find the value of   **i** w,   **ii** x,   **iii** y.

b   Work out the area of the big trapezium.

Exercise 33C

1 Triangles **P** and **Q** are similar.
The area of triangle **P** is 10 cm².

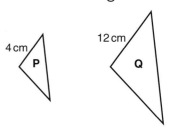

Calculate the area of triangle **Q**.

2 Circles **P** and **Q** are similar.
The diameter of circle **Q** is 4 times the diameter of circle **P**.
The area of circle **Q** is 48 cm².

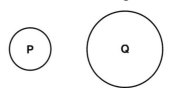

Calculate the area of circle **P**.

3 Cylinders **P** and **Q** are similar.
The surface area of cylinder **P** is 36 cm².

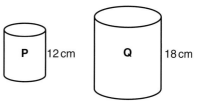

Calculate the surface area of cylinder **Q**.

4 Cones **P** and **Q** are similar.
The surface area of cone **Q** is 25 times the surface area of cone **P**.

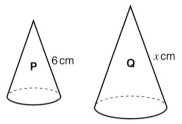

Calculate the value of x.

5 Pyramids **P** and **Q** are similar.
The surface area of pyramid **P** is 200 cm² and the area of pyramid **Q** is 512 cm².

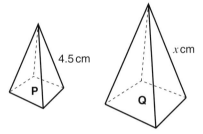

Calculate the value of x.

Exercise 33D

1 Pyramids **P** and **Q** are similar.
The volume of pyramid **P** is 12 cm³.

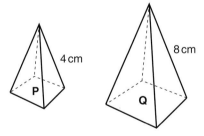

Calculate the volume of pyramid **Q**.

2 Cones **P** and **Q** are similar.
The volume of cone **Q** is 80 cm³.

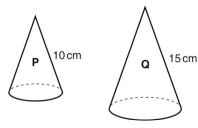

Calculate the volume of cone **P**.

3 Cuboids **P** and **Q** are similar.
The volume of cuboid **P** is 64 times the volume of cuboid **Q**.

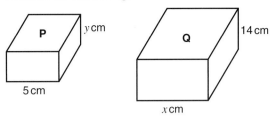

Calculate the value of   **a** x **b** y.

4 Pyramids **P** and **Q** are similar.
The volume of pyramid **P** is 48 cm³.
The volume of pyramid **Q** is 750 cm³.

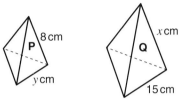

Calculate the value of   **a** x **b** y.

Exercise 33E

1 Pyramids **P** and **Q** are similar.
The surface area of pyramid **Q** is 16 times the surface area of pyramid **P**.
The volume of pyramid **P** is 15 cm³.
Calculate the volume of pyramid **Q**.

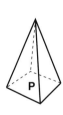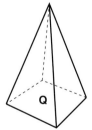

2 Prisms **P** and **Q** are similar.
The surface area of prism **P** is 32 cm².
The surface area of prism **Q** is 72 cm².
The volume of prism **P** is 16 cm³.

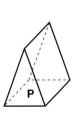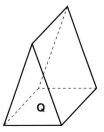

Calculate the volume of prism **Q**.

3 **X** and **Y** are two geometrically similar solid shapes.
The total area of shape **X** is 450 cm².
The total area of shape **Y** is 800 cm².
The volume of shape **X** is 1350 cm³.
Calculate the volume of shape **Y**.
(1387 November 2004)

4 Two similar boxes have volumes of 2000 cm³ and 16 000 cm³.
The area of the base of the larger box is 60 cm².
Calculate the area of the base, in cm², of the smaller box. *(1384 June 1996)*

5

 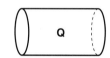

Diagram **NOT** accuratey drawn

4 cm

Two cylinders, **P** and **Q**, are mathematically similar.
The total surface area of cylinder **P** is 90π cm².
The total surface area of cylinder **Q** is 810π cm².
The length of cylinder **P** is 4 cm.

a Work out the length of cylinder **Q**.

The volume of cylinder **P** is 100π cm³.

b Work out the volume of cylinder **Q**.
Give your answer as a multiple of π.
(1387 June 2005)

Chapter 34 Proportion

Exercise 34A

1 y is directly proportional to x.

  a $y = 12$ when $x = 2$. Find y when $x = 8$
  b $y = 8$ when $x = 4$. Find y when $x = 7$
  c $y = 12$ when $x = 3$. Find y when $x = 8$
  d $y = 12$ when $x = 5$. Find y when $x = 15$

2 y is directly proportional to x.

  a $y = 10$ when $x = 5$. Find x when $y = 30$
  b $y = 18$ when $x = 6$. Find x when $y = 24$
  c $y = 8$ when $x = 5$. Find x when $y = 28$
  d $y = 10$ when $x = 8$. Find x when $y = 25$

3 The weight W grams of a pile of paper is directly proportional to the number of sheets, N, in the pile.
When $N = 220, W = 40$

  a Find a formula for W in terms of N.
  b Find the value of W when $N = 330$.
  c Find the value of N when $W = 100$.

4 The number of words, N, typed by a typist is directly proportional to the time, T, seconds she takes.
When $T = 40, N = 60$

  a Find a formula for N in terms of T.
  b Find the value of M when $T = 260$.
  c Find the value of T when $N = 120$.

5 The distance d, metres, a car travels in 10 seconds is directly proportional to the speed, s kilometres, of the car.
When $s = 54, d = 150$.

  a Find a formula for d in terms of s.
  b Find the value of d when $s = 72$.
  c Find the value of s when $d = 250$.

Exercise 34B

 1 y is proportional to the square of x.

  a When $x = 4, y = 32$. Find y when $x = 3$
  b When $x = 2, y = 12$. Find y when $x = 7$
  c When $x = 2, y = 10$. Find y when $x = 6$
  d When $x = 5, y = 200$. Find y when $x = 4$

2 y is proportional to the square of x.

  a When $x = 3, y = 18$. Find x when $y = 50$
  b When $x = 2, y = 20$.
    Find x when $y = 125$
  c When $x = 4, y = 48$.
    Find x when $y = 243$
  d When $x = 2, y = 10$.
    Find x when $y = 250$

 3 The area, A cm², of the screen of a set of televisions of different sizes is proportional to the square of the length, L cm, of the diagonal.
When $L = 20$ $A = 160$

a Find a formula for A in terms of L.

b Find the value of A when $L = 30$

c Find the value of L when $A = 90$

4 The heat, H calories, given out by an electric fire each second is proportional to the square of the current, I amps, passing through the fire.
When $I = 5, H = 300$

   a Find a formula for H in terms of I.

   b Find the value of H when $I = 6$.

   c Find the value of I when $H = 48$.

5 y is proportional to the cube of x.

   a When $x = 1, y = 8$. Find y when $x = 2$.

   b When $x = 2, y = 24$. Find y when $x = 3$.

   c When $x = 5, y = 25$. Find y when $x = 10$.

   d When $x = 10, y = 800$.
   Find y when $x = 5$

6 The mass, M kg, of an iron sphere is proportional to the cube of the radius r cm of the sphere.
When $r = 10, M = 40$

   a Find a formula for M in terms of r.

   b Find M when $r = 20$.

   c Find r when $M = 1080$.

Exercise 34C

1 y is inversely proportional to x.

   a $y = 8$ when $x = 4$. Find y when $x = 8$.

   b $y = 10$ when $x = 2$. Find y when $x = 32$.

   c $y = 15$ when $x = 10$. Find y when $x = 6$.

   d $y = 15$ when $x = 8$. Find y when $x = 5$.

2 The time, T seconds, it takes a car to travel a fixed distance is inversely proportional to its speed, S kilometres per hour.
When $S = 54, T = 20$

   a Find a formula for T in terms of S.

   b Find T when $S = 72$.

   c Find S when $T = 12$.

3 The wavelength, L m, of a sound wave is inversely proportional to its frequency, f Hertz.
When $f = 450, L = 2$

   a Find a formula for L in terms of f.

   b Find L when $f = 90$.

   c Find f when $L = 135$.

4 The time taken, T minutes, to word process a document is inversely proportional to the speed of typing, W words per minute.
When $W = 60, T = 20$

   a Find a formula for T in terms of W.

   b Find T when $W = 45$.

   c Find W when $T = 90$.

5 y is inversely proportional to the square of x.

   a $y = 4$ when $x = 2$. Find y when $x = 1$.

   b $y = 2$ when $x = 3$. Find y when $x = 2$.

   c $y = 10$ when $x = 5$. Find y when $x = 10$.

   d $y = 3$ when $x = \frac{1}{2}$. Find y when $x = \frac{3}{4}$.

6 The gravitational force F on a satellite round the Earth is inversely proportional to the square of the distance, d m, of the satellite from the centre of the earth.
When $d = 10^7, F = 12.8$

   a Find a formula for F in terms of d.

   b Find the value of r for which $F = 20$

7 The rate of evaporation E cm^3 per second of a water droplet is inversely proportional to the square of the radius r cm of the droplet.
When $r = 2, E = 0.09$

   a Find a formula for E in terms of r.

   b Find E when $r = 0.5$

   c Find r when $E = 0.25$

8 The force of attraction, F, between 2 magnets is inversely proportional to the distance, D centimetres, between them.
When $D = 6, F = 4$
Find F when $D = 8$.

9 The sound intensity I, at a distance x metres from a sound source is inversely proportional to the square of x.

When $x = 20$, $I = 14$

Find x when $I = 2$. Give your answer correct to 3 significant figures.

Exercise 34D

1 y is directly proportional to the square root of x.

a $y = 8$ when $x = 4$.
Find y when $x = 16$

b $y = 6$ when $x = 4$.
Find y when $x = 36$

c $y = 50$ when $x = 25$.
Find y when $x = 100$

d $y = 60$ when $x = 25$.
Find y when $x = 400$

2 y is inversely proportional to the square root of x.

a $y = 1$ when $x = 16$. Find y when $x = 25$
b $y = 3$ when $x = 25$. Find y when $x = 36$
c $y = 9$ when $x = 81$. Find y when $x = 9$
d $y = \frac{1}{2}$ when $x = 16$. Find y when $x = 25$

3 The time, T seconds, taken by a car to accelerate to a given speed is directly proportional to the square root of the distance covered d metres

When $d = 100$, $T = 5$

a Find a formula for T in terms of d.
b Find the value of T when $d = 225$
c Find the value of d when $T = 8$

4 The number of vibrations per second, n, of a spring is inversely proportional to the unstretched length L cm, of the spring.

When $L = 100$, $n = 2$

a Find a formula for n in terms of L
b Find the value of n when $L = 25$

A spring does 20 vibrations in 8 seconds.

c Find its unstretched length

Chapter 35 Vectors

Exercise 35A

1 Draw accurately and label the following vectors

i Vector **a** with magnitude 4 cm and direction north.

ii Vector **b** with magnitude 3 cm and direction with bearing 070°.

iii Vector **c** with magnitude 2.5 cm and direction with bearing 310°.

iv Vector $\overrightarrow{AB}$ with magnitude 5 cm and direction with bearing 270°.

v Vector $\overrightarrow{MN}$ with magnitude 6 cm and direction with bearing 120°.

2 On squared paper, draw and label the following vectors

i $\mathbf{a} = \begin{pmatrix} 3 \\ 0 \end{pmatrix}$ **ii** $\mathbf{b} = \begin{pmatrix} 2 \\ 5 \end{pmatrix}$

iii $\mathbf{c} = \begin{pmatrix} -4 \\ 3 \end{pmatrix}$ **iv** $\overrightarrow{PQ} = \begin{pmatrix} -1 \\ -6 \end{pmatrix}$

v $\overrightarrow{CD} = \begin{pmatrix} 0 \\ -2 \end{pmatrix}$ **vi** $\overrightarrow{EF} = \begin{pmatrix} 1 \\ -3 \end{pmatrix}$

3 a On graph paper, mark and label the following points

$A(1, 3)$, $B(5, 8)$, $C(-3, 4)$, $D(4, -5)$, $E(-4, -3)$, $F(5, -5)$.

b Write as column vectors

i $\overrightarrow{AB}$ **ii** $\overrightarrow{CD}$ **iii** $\overrightarrow{DE}$

iv $\overrightarrow{BF}$ **v** $\overrightarrow{BA}$ **vi** $\overrightarrow{FC}$

4 A is the point with coordinates $(2, 3)$.

$\overrightarrow{AB} = \begin{pmatrix} 5 \\ -4 \end{pmatrix}$.

Find the coordinates of B.

(4400 November 2005)

5 A is the point $(2, 3)$ and B is the point $(-2, 0)$

a Find $\overrightarrow{AB}$ as a column vector.

C is the point such that $\overrightarrow{BC} = \begin{pmatrix} 4 \\ 9 \end{pmatrix}$

b Write down the coordinates of the point C.

X is the midpoint of AB. O is the origin.

c   Find $\overrightarrow{OX}$ as a column vector.

(1385 June 1995)

Exercise 35B

1 $ABCDEF$ is a regular
   hexagon with centre O.

a   Write down three vectors
     in the diagram that
     are equal to $\overrightarrow{AB}$.

b   Explain why $\overrightarrow{CB} = \overrightarrow{EF}$.

2 Work out the magnitude of each of these
   vectors. Where necessary, answers may be
   left as surds.

i $\mathbf{a} = \begin{pmatrix} 8 \\ 6 \end{pmatrix}$ ii $\mathbf{b} = \begin{pmatrix} 2.5 \\ 6 \end{pmatrix}$

iii $\mathbf{c} = \begin{pmatrix} 24 \\ -7 \end{pmatrix}$ iv $\mathbf{d} = \begin{pmatrix} 6 \\ 4 \end{pmatrix}$

v $\mathbf{e} = \begin{pmatrix} -3 \\ -8 \end{pmatrix}$ vi $\overrightarrow{AB} = \begin{pmatrix} 35 \\ 12 \end{pmatrix}$

vii $\overrightarrow{MN} = \begin{pmatrix} -4 \\ 5 \end{pmatrix}$

3 The diagram is a sketch.
   P is the point $(2, 3)$ Q is the point $(6, 6)$

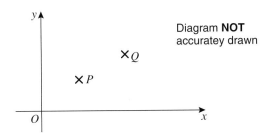

Diagram **NOT**
accuratey drawn

a   Write down the vector $\overrightarrow{PQ}$

Write your answer as a column vector $\begin{pmatrix} x \\ y \end{pmatrix}$

$PQRS$ is a parallelogram. $\overrightarrow{PR} = \begin{pmatrix} 4 \\ 7 \end{pmatrix}$.

b   Find the vector $\overrightarrow{QS}$.

Write your answer as a column vector $\begin{pmatrix} x \\ y \end{pmatrix}$

(1387 June 2006)

Exercise 35C

1 The diagram shows three vectors **a**, **b** and **c**.

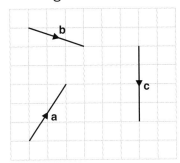

On a copy of this diagram draw the vector
i   **a** + **b**,   ii   **b** + **c**,   iii   **(a + b)** + **c**

2 Work out

a $\begin{pmatrix} 3 \\ 1 \end{pmatrix} + \begin{pmatrix} 4 \\ 7 \end{pmatrix}$ b $\begin{pmatrix} 2 \\ 8 \end{pmatrix} + \begin{pmatrix} 6 \\ -5 \end{pmatrix}$

c $\begin{pmatrix} 3 \\ -4 \end{pmatrix} + \begin{pmatrix} -2 \\ -5 \end{pmatrix}$ d $\begin{pmatrix} 0 \\ 4 \end{pmatrix} + \begin{pmatrix} -4 \\ 3 \end{pmatrix}$

e $\begin{pmatrix} 6 \\ -3 \end{pmatrix} + \begin{pmatrix} -6 \\ 3 \end{pmatrix}$

3 $ABCD$ is a quadrilateral, where

$\overrightarrow{AB} = \begin{pmatrix} 2 \\ 4 \end{pmatrix}$, $\overrightarrow{BC} = \begin{pmatrix} 3 \\ 2 \end{pmatrix}$ and $\overrightarrow{CD} = \begin{pmatrix} 4 \\ -8 \end{pmatrix}$.

a   Find, as a column vector, i $\overrightarrow{AC}$, ii $\overrightarrow{AD}$.

b   i   Work out $\overrightarrow{AB} + \overrightarrow{BC} + \overrightarrow{CD}$ as a
         column vector.

   ii   Explain why it must be true that
         $\overrightarrow{AB} + \overrightarrow{BC} + \overrightarrow{CD} = \overrightarrow{AD}$

The quadrilateral is on a centimetre grid.

c   Work out the length of each side of
     quadrilateral $ABCD$. Give each length
     correct to 1 decimal place.

Exercise 35D

1 The diagram shows
   two vectors **a** and **b**.
   On squared paper
   draw the vectors

i   **3a**   ii $\frac{1}{2}$**a**

iii   **−b**   iv   **−2b**

v   **a − b**

vi   **2a + b**

vii   **a − 2b**

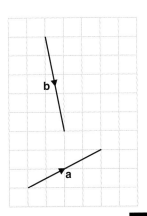

2 $a = \begin{pmatrix} 3 \\ -5 \end{pmatrix}$ $b = \begin{pmatrix} -2 \\ 4 \end{pmatrix}$ $c = \begin{pmatrix} 5 \\ 8 \end{pmatrix}$

a Find, as a column vector,

    i $4a$ **ii** $-\frac{1}{2}b$

    iii $5a + 2c$ **iv** $a + 2b - 4c$

    v $\frac{1}{4}(2c - 3b)$

b Find, as a surd, the magnitude of the vector $4b + 3a$

3 The points M and N have coordinates $(3, 1)$ and $(7, 11)$

a Write down as a column vector the position vector of

    i the point M,   **ii** the point N.

The point P is the midpoint of MN.

b Find as a column vector the position vector of P.

c Using column vectors, show that $\overrightarrow{MP} = \frac{1}{2}\overrightarrow{MN}$

4 Simplify

    i $4m + 7n + 2m - 8n$

    ii $5(2a - 4b) - 3(3a - 5b)$

    iii $2p + 3q + \frac{1}{2}(4p - 7q)$

5 $PQRS$ is a trapezium.

QP is parallel to RS

$QP = 3RS$

$\overrightarrow{QR} = a$, $\overrightarrow{RS} = c$

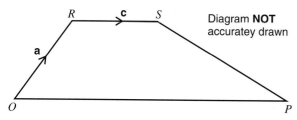

Diagram **NOT** accuratey drawn

Express in terms of a and/or c

    i $\overrightarrow{QP}$ **ii** $\overrightarrow{SP}$

(1388 November 2005)

6 $ABCD$ is a straight line.

O is a point so that $\overrightarrow{OA} = a$ and $\overrightarrow{OB} = b$

B is the midpoint of AC.

C is the midpoint of AD.

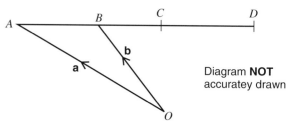

Diagram **NOT** accuratey drawn

Express, in terms of a and b, the vectors

    i $\overrightarrow{AC}$ **ii** $\overrightarrow{OD}$

(1387 November 2005)

Exercise 35E

1 $OABC$ is a trapezium.

OC is parallel to AB.

$\overrightarrow{OA} = a$, $\overrightarrow{OC} = c$

$AB = 2OC$

X is the point on AB such that

$AX : XB = 3 : 1$

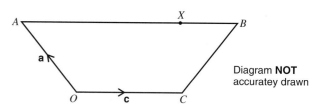

Diagram **NOT** accuratey drawn

Express $\overrightarrow{XC}$ in terms of a and c.

(1388 March 2003)

2 OPQ is a triangle.

T is the point on PQ for which

$PT : TQ = 2 : 1$

$\overrightarrow{OP} = a$ and $\overrightarrow{OQ} = b$

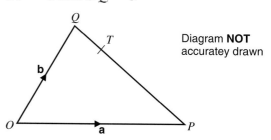

Diagram **NOT** accuratey drawn

a Write down, in terms of a and b, an expression for $\overrightarrow{PQ}$.

b Express $\overrightarrow{OT}$ in terms of a and b.

Give your answer in its simplest form.

(1387 November 2003)

3 In the quadrilateral $OPQR$

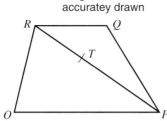

Diagram **NOT** accuratey drawn

$\overrightarrow{OP} = \mathbf{a}$

$\overrightarrow{OQ} = \mathbf{b}$

$\overrightarrow{OR} = \mathbf{b} - \frac{3}{4}\mathbf{a}$

a **i** Find, in terms of $\mathbf{a}$, the vector $\overrightarrow{RQ}$.
  ii Explain what this result shows about the side RQ of the quadrilateral.
b Express $\overrightarrow{PR}$ in terms of $\mathbf{a}$ and $\mathbf{b}$.
The point T lies on PR such that $\overrightarrow{PT} = \frac{4}{7}\overrightarrow{PR}$.
c **i** Find an expression, in terms of $\mathbf{a}$ and $\mathbf{b}$, for $\overrightarrow{OT}$.
  ii Explain what this result shows about the position of T.

4 $ABCD$ is a parallelogram. The diagonals of the parallelogram intersect at O.
$\overrightarrow{OA} = \mathbf{a}$ $\overrightarrow{OB} = \mathbf{b}$

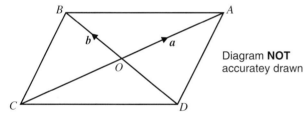

Diagram **NOT** accuratey drawn

a Write down an expression, in terms of $\mathbf{a}$ and $\mathbf{b}$, for
  i $\overrightarrow{CA}$ **ii** $\overrightarrow{BA}$ **iii** $\overrightarrow{AD}$
X is the point such that $\overrightarrow{OX} = 2\mathbf{a} - \mathbf{b}$
b **i** Write down an expression, in terms of $\mathbf{a}$ and $\mathbf{b}$, for $\overrightarrow{AX}$.
  ii Explain why B, A and X lie on the same straight line.

(1385 November 1997)

Chapter 36 Transformation of functions

Exercise 36A

1 $f(x) = 4x + 3$ Find the value of
  a $f(2)$ **b** $f(5)$ **c** $f(-3)$
  d $f(\frac{1}{2})$ **e** $f(-\frac{3}{4})$ **f** $f(0)$

2 $g(x) = 2 - x^2$ Find the value of
  a $g(1)$ **b** $g(3)$ **c** $g(-5)$
  d $g(0)$ **e** $g(\frac{1}{2})$ **f** $g(\frac{3}{4})$

3 $f(x) = x^3$ Find an expression for
  a $f(x) + 2$ **b** $f(x) - 5$ **c** $3f(x)$
  d $\frac{1}{2}f(x)$ **e** $-f(x)$ **f** $2f(x) - 4$

4 $f(x) = x^3$ Find an expression for
  a $f(x + 2)$ **b** $f(x - 5)$ **c** $f(3x)$
  d $f(\frac{1}{2}x)$ **e** $f(-x)$ **f** $f(2x - 4)$

5 $g(x) = 1 - 2x$ Find an expression for
  a $g(x) + 5$ **b** $g(x - 2)$ **c** $g(2x)$
  d $3g(x)$ **e** $-g(x)$ **f** $g(-x)$

Exercise 36B

1 The equation of one of the graphs is given. Write down the equation of each of the other two graphs.

i

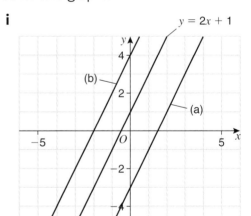

$y = 2x + 1$

ii

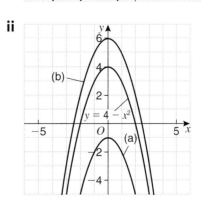

$y = 4 - x^2$

2 a The graph of $y = x$ is translated by 8 units vertically in the positive y direction. Write down the equation of the new graph.

b The graph of $y = x$ is translated by 5 units vertically in the negative y direction. Write down the equation of the new graph.

c The graph of $y = x^2$ is translated by 6 units vertically in the negative y direction. Write down the equation of the new graph.

d The graph of $y = x^2$ is translated by 3 units vertically in the positive y direction. Write down the equation of the new graph.

e Describe the transformation that will map the graph of $y = x^2$ onto the graph of $y = x^2 - 16$

3 The graph of $y = f(x)$ is shown. Copy and sketch, on the same axes, the graph of

a $y = f(x) - 2$

b $y = f(x) + 3$

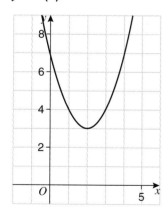

Exercise 36C

1 The equation of one of the graphs is given. Write down the equation of each of the other two graphs.

i

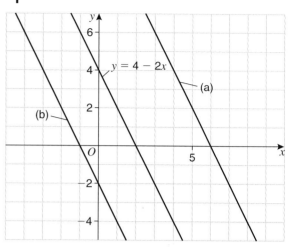

ii

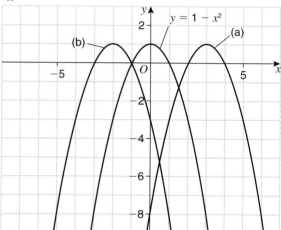

2 The graph of $y = f(x)$ is shown. Copy and sketch, on the same axes, the graph of

a $y = f(x + 2)$ **b** $y = f(x - 5)$

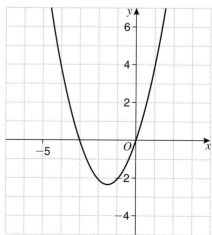

3 The graph of $y = f(x)$ is shown.

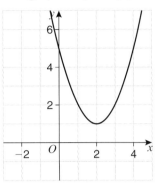

Sketch the graph of

a $y = f(x + 3) - 2$ **b** $y = f(x - 3) + 2$

Exercise 36D

1 Copy the sketch of each graph and draw on the axes the graph of

i $y = -f(x)$ **ii** $y = f(-x)$

a

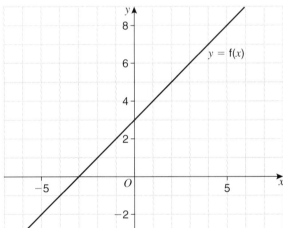

b

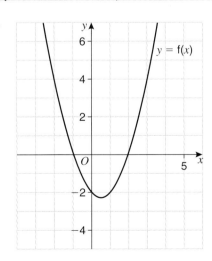

c

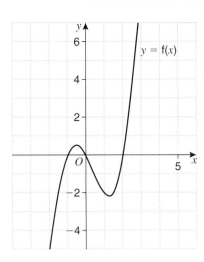

2 The graph of $y = 5 - 2x$ is reflected in the x-axis. Write down the equation of the new graph.

3 The graph of $y = 3x - 2$ is reflected in the y-axis. Write down the equation of the new graph.

4 The graph of $y = f(x)$ has a vertex at $(2, -5)$. Write down the coordinates of the vertex of

a $y = -f(x)$ **b** $y = f(-x)$

5 This is a sketch of the curve with equation $y = f(x)$.

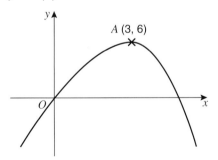

The only maximum point of the curve $y = f(x)$ is $A(3, 6)$.

Write down the coordinates of the maximum point for curves with each of the following equations.

i $y = f(x + 2)$

ii $y = f(x) + 4$

iii $y = f(-x)$

(1385 May 2002)

Exercise 36E

1 Copy the sketch of the graph and draw on the same axes the graphs of

   a $y = 2f(x)$ **b** $y = \frac{1}{3}f(x)$

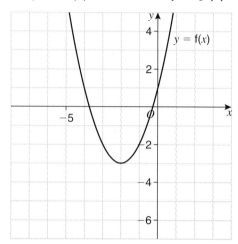

2 Copy the sketch of the graph and draw on the same axes the graphs of

   a $y = f(\frac{1}{3}x)$ **b** $y = f(2x)$

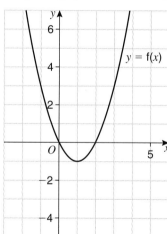

3 The graph of $y = 2x - 1$ is stretched by a scale factor of 4 parallel to the y-axis. Write down the equation of the new graph.

4 The graph of $y = x(x + 1)$ is stretched by a scale factor of $\frac{1}{3}$ parallel to the x-axis. Write down the equation of the new graph.

5 The graph of $y = f(x)$ where $f(x) = 2x^2$ is stretched. Write down the equation of the new graph when the stretch is

   a   parallel to the y-axis with a scale factor of $\frac{1}{2}$

   b   parallel to the x-axis with a scale factor of $\frac{1}{2}$

6 The graph of $y = f(x)$ has a vertex at $(9, -12)$. Write down the coordinates of the vertex of

   a $y = \frac{1}{2}f(x)$ **b** $y = f(\frac{1}{2}x)$

   c $y = \frac{1}{3}f(x)$ **d** $y = f(3x)$

Exercise 36F

1 **a**   Sketch the graph $y = \sin x°$ for $0 \leqslant x \leqslant 360$

   b   On the same set of axes sketch the graphs of

      i $y = 2 + \sin x°$ **ii** $y = 3 \sin x°$

2 **a**   Sketch the graph of $y = \cos x°$ for $0 \leqslant x \leqslant 360$

   b   On the same set of axes sketch the graphs of

      i $y = 2 \cos x°$ **ii** $y = \cos 2x°$

3 The minimum possible value of $\cos x°$ is -1 Write down the minimum possible value of

   a $3 \cos x°$ **b** $2 + \cos x°$

   c $\cos 2x°$ **d** $\cos x° - 5$

   e $\frac{1}{2} \cos x°$ **f** $3 \cos x° - 2$

4

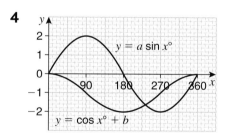

The diagram shows part of two graphs.

The equation of one graph is $y = a \sin x°$
The equation of the other graph is $y = \cos x° + b$

   a   Use the graphs to find the value of a and the value of b.

   b   Use the graphs to find the values of x in the range $0 \leqslant x \leqslant 720$ when $a \sin x° = \cos x° + b$.

   c   Use the graphs to find the value of $a \sin x° - (\cos x° + b)$ when $x = 450$.

(1387 November 2004)